Android™

A Programmer's Guide

J.F. DiMarzio

New York Chicago San Francisco
Lisbon London Madrid Mexico City
Milan New Delhi San Juan
Seoul Singapore Sydney Toronto

The McGraw·Hill Companies

Cataloging-in-Publication Data is on file with the Library of Congress

McGraw-Hill books are available at special quantity discounts to use as premiums and sales promotions, or for use in corporate training programs. To contact a special sales representative, please visit the Contact Us page at www.mhprofessional.com.

Android™: A Programmer's Guide

1234567890 DOC DOC 0198

ISBN 978-0-07-159988-7
MHID 0-07-159988-6

Sponsoring Editor Roger Stewart
Project Editor Janet Walden
Acquisitions Coordinator Carly Stapleton
Technical Editor Gilbert L. Polo
Copy Editor Bill McManus
Proofreader Francesca Ferrie
Indexer Claire Splan
Production Supervisor George Anderson
Composition Apollo Publishing Services
Illustration Apollo Publishing Services
Art Director, Cover Jeff Weeks

Android™

A Programmer's Guide

This book is dedicated to Suzannah, Christian, and Sophia

About the Author

J.F. DiMarzio is a developer with over 15 years of experience in networking and application development and is the author of seven books on computing technologies. He has become a leading resource in the fields of IT consulting and development. He lives in Central Florida.

About the Technical Editor

Gilbert L. Polo is a software developer with over 20 years of experience working in the telecommunications, financial, and, most recently, educational industries. He has programmed in various languages including C, C++, Java, and C#.

Contents at a Glance

Contents

Acknowledgments

I would like to thank everyone who participated in the creation of this book. My agent, Neil Salkind; Roger, Carly, Janet, Bill, and the crew at McGraw-Hill; Gil Polo; and everyone at Studio B.

I would also like to thank my family, Suzannah, Christian, and Sophia; Brett, Robert, Roger, Zack, Mark, Kurt, Walter, Walter, Walter, Steve, Steve, Steve, and Gary—and all my colleagues in Central Florida; and anyone else whom I may have forgotten.

Introduction

Welcome to *Android: A Programmer's Guide*. This book has been designed to give you the best first step toward the exciting new frontier of open source mobile development. Android is the newest mobile device operating system, and this is one of the first books to help the average programmer become a fearless Android developer. Through the course of this book, you will be introduced to the fundamentals of mobile device application development using the Open Handset Alliance's Android platform. By the end of this book, you will be able to confidently create your own mobile device programs.

The format of this book is such that it will take you through Android application development in a logical manner. The book begins by examining the architecture of Android as a platform, looking at how it was developed, what it can run on, and what tools are required to develop programs for it. After discussing and installing the development tools, Android SDK, and the Eclipse development environment (Chapters 2, 3, and 4), the book dives directly into designing and creating Android applications (Chapter 5). The book concludes with instructions on tying your applications to existing Google tools such as Google Maps (Chapters 9 and 11) and GTalk (Chapter 10). A quick reference guide is also included in Chapter 12.

This book is a programmer's guide, not a beginner's guide, meaning that you do need to possess some programming skills to get the most from it. Foremost among these skills is a working knowledge of Java programming fundamentals. Android applications are developed in Java and run on the Linux 2.6 kernel. If you are a quick learner, you may be able to understand what is going on with just some basic object-oriented programming (OOP) experience. Chapter 2 explains how to download and install the preferred integrated development environment, Eclipse. All the code samples and screenshots in this book are provided using Eclipse (Europa release) and the Android plugin for Eclipse.

Any comments, questions, or suggestions about any of the material in this book can be forwarded directly to the author at **jfdimarzio@jfdimarzio.com**.

Chapter 1

What Is Android?

Key Skills & Concepts

- History of embedded device programming
- Explanation of Open Handset Alliance
- First look at the Android home screen

It can be said that, for a while, traditional desktop application developers have been spoiled. This is not to say that traditional desktop application development is easier than other forms of development. However, as traditional desktop application developers, we have had the ability to create almost any kind of application we can imagine. I am including myself in this grouping because I got my start in desktop programming.

One aspect that has made desktop programming more accessible is that we have had the ability to interact with the desktop operating system, and thus interact with any underlying hardware, pretty freely (or at least with minimal exceptions). This kind of freedom to program independently, however, has never really been available to the small group of programmers who dared to venture into the murky waters of cell phone development.

NOTE

I refer to two different kinds of developers in this discussion: *traditional desktop application developers*, who work in almost any language and whose end product, applications, are built to run on any "desktop" operating system; and *Android developers*, Java developers who develop for the Android platform. This is not for the purposes of saying one is by any means better or worse than the other. Rather, the distinction is made for purposes of comparing the development styles and tools of desktop operating system environments to the mobile operating system environment, Android.

Brief History of Embedded Device Programming

For a long time, cell phone developers comprised a small sect of a slightly larger group of developers known as embedded device developers. Seen as a less "glamorous" sibling to desktop—and later web—development, embedded device development typically got the proverbial short end of the stick as far as hardware and operating system features, because embedded device manufacturers were notoriously stingy on feature support. Embedded device manufacturers typically needed to guard their hardware secrets closely, so they gave embedded device developers few libraries to call when trying to interact with a specific device.

Embedded devices differ from desktops in that an embedded device is typically a "computer on a chip." For example, consider your standard television remote control; it is not really seen as an overwhelming achievement of technological complexity. When any button is pressed, a chip interprets the signal in a way that has been programmed into the device. This allows the device to know what to expect from the input device (key pad), and how to respond to those commands (for example, turn on the television). This is a simple form of embedded device programming. However, believe it or not, simple devices such as these are definitely related to the roots of early cell phone devices and development.

Most embedded devices ran (and in some cases still run) proprietary operating systems. The reason for choosing to create a proprietary operating system rather than use any consumer system was really a product of necessity. Simple devices did not need very robust and optimized operating systems.

As a product of device evolution, many of the more complex embedded devices, such as early PDAs, household security systems, and GPSs, moved to somewhat standardized operating system platforms about five years ago. Small-footprint operating systems such as Linux, or even an embedded version of Microsoft Windows, have become more prevalent on many embedded devices. Around this time in device evolution, cell phones branched from other embedded devices onto their own path. This branching is evident when you examine their architecture.

Nearly since their inception, cell phones have been fringe devices insofar as they run on proprietary software—software that is owned and controlled by the manufacturer, and is almost always considered to be a "closed" system. The practice of manufacturers using proprietary operating systems began more out of necessity than any other reason. That is, cell phone manufacturers typically used hardware that was completely developed in-house, or at least hardware that was specifically developed for the purposes of running cell phone equipment. As a result, there were no openly available, off-the-shelf software packages or solutions that would reliably interact with their hardware. Since the manufacturers also wanted to guard very closely their hardware trade secrets, some of which could be revealed by allowing access to the software level of the device, the common practice was, and in most cases still is, to use completely proprietary and closed software to run their devices. The downside to this is that anyone who wanted to develop applications for cell phones needed to have intimate knowledge of the proprietary environment within which it was to run. The solution was to purchase expensive development tools directly from the manufacturer. This isolated many of the "homebrew" developers.

NOTE

A growing culture of homebrew developers has embraced cell phone application development. The term "homebrew" refers to the fact that these developers typically do not work for a cell phone development company and generally produce small, one-off products on their own time.

Another, more compelling "necessity" that kept cell phone development out of the hands of the everyday developer was the hardware manufacturers' solution to the "memory versus need" dilemma. Until recently, cell phones did little more than execute and receive phone calls, track your contacts, and possibly send and receive short text messages; not really the "Swiss army knives" of technology they are today. Even as late as 2002, cell phones with cameras were not commonly found in the hands of consumers.

By 1997, small applications such as calculators and games (Tetris, for example) crept their way onto cell phones, but the overwhelming function was still that of a phone dialer itself. Cell phones had not yet become the multiuse, multifunction personal tools they are today. No one yet saw the need for Internet browsing, MP3 playing, or any of the multitudes of functions we are accustomed to using today. It is possible that the cell phone manufacturers of 1997 did not fully perceive the need consumers would have for an all-in-one device. However, even if the need was present, a lack of device memory and storage capacity was an even bigger obstacle to overcome. More people may have wanted their devices to be all-in-one tools, but manufacturers still had to climb the memory hurdle.

To put the problem simply, it takes memory to store and run applications on any device, cell phones included. Cell phones, as a device, until recently did not have the amount of memory available to them that would facilitate the inclusion of "extra" programs. Within the last two years, the price of memory has reached very low levels. Device manufacturers now have the ability to include more memory at lower prices. Many cell phones now have more standard memory than the average PC had in the mid-1990s. So, now that we have the need, and the memory, we can all jump in and develop cool applications for cell phones around the world, right? Not exactly.

Device manufacturers still closely guard the operating systems that run on their devices. While a few have opened up to the point where they will allow some Java-based applications to run within a small environment on the phone, many do not allow this. Even the systems that do allow some Java apps to run do not allow the kind of access to the "core" system that standard desktop developers are accustomed to having.

Open Handset Alliance and Android

This barrier to application development began to crumble in November of 2007 when Google, under the Open Handset Alliance, released Android. The Open Handset Alliance is a group of hardware and software developers, including Google, NTT DoCoMo, Sprint Nextel, and HTC, whose goal is to create a more open cell phone environment. The first product to be released under the alliance is the mobile device operating system, Android. (For more information about the Open Handset Alliance, see www.openhandsetalliance.com.)

With the release of Android, Google made available a host of development tools and tutorials to aid would-be developers onto the new system. Help files, the platform software development kit (SDK), and even a developers' community can be found at Google's Android website, http://code.google.com/android. This site should be your starting point, and I highly encourage you to visit the site.

NOTE

Google, in promoting the new Android operating system, even went as far as to create a $10 million contest looking for new and exciting Android applications.

While cell phones running Linux, Windows, and even PalmOS are easy to find, as of this writing, no hardware platforms have been announced for Android to run on. HTC, LG Electronics, Motorola, and Samsung are members of the Open Handset Alliance, under which Android has been released, so we can only hope that they have plans for a few Android-based devices in the near future. With its release in November 2007, the system itself is still in a software-only beta. This is good news for developers because it gives us a rare advance look at a future system and a chance to begin developing applications that will run as soon as the hardware is released.

NOTE

This strategy clearly gives the Open Handset Alliance a big advantage over other cell phone operating system developers, because there could be an uncountable number of applications available immediately for the first devices released to run Android.

Introduction to Android

Android, as a system, is a Java-based operating system that runs on the Linux 2.6 kernel. The system is very lightweight and full featured. Figure 1-1 shows the unmodified Android home screen.

Figure 1-1 The current Android home screen as seen on the Android Emulator.

Android applications are developed using Java and can be ported rather easily to the new platform. If you have not yet downloaded Java or are unsure about which version you need, I detail the installation of the development environment in Chapter 2. Other features of Android include an accelerated 3-D graphics engine (based on hardware support), database support powered by SQLite, and an integrated web browser.

If you are familiar with Java programming or are an OOP developer of any sort, you are likely used to programmatic user interface (UI) development—that is, UI placement which is handled directly within the program code. Android, while recognizing and allowing for programmatic UI development, also supports the newer, XML-based UI layout. XML UI layout is a fairly new concept to the average desktop developer. I will cover both the XML UI layout and the programmatic UI development in the supporting chapters of this book.

One of the more exciting and compelling features of Android is that, because of its architecture, third-party applications—including those that are "home grown"—are executed with the same system priority as those that are bundled with the core system. This is a major departure from most systems, which give embedded system apps a greater execution priority than the thread priority available to apps created by third-party developers. Also, each application is executed within its own thread using a very lightweight virtual machine.

Aside from the very generous SDK and the well-formed libraries that are available to us to develop with, the most exciting feature for Android developers is that we now have access to anything the operating system has access to. In other words, if you want to create an application that dials the phone, you have access to the phone's dialer; if you want to create an application that utilizes the phone's internal GPS (if equipped), you have access to it. The potential for developers to create dynamic and intriguing applications is now wide open.

On top of all the features that are available from the Android side of the equation, Google has thrown in some very tantalizing features of its own. Developers of Android applications will be able to tie their applications into existing Google offerings such as Google Maps and the omnipresent Google Search. Suppose you want to write an application that pulls up a Google map of where an incoming call is emanating from, or you want to be able to store common search results with your contacts; the doors of possibility have been flung wide open with Android.

Chapter 2 begins your journey to Android development. You will learn the hows and whys of using specific development environments or integrated development environments (IDE), and you will download and install the Java IDE Eclipse.

Ask the Expert

Q: What is the difference between Google and the Open Handset Alliance?

A: Google is a member of the Open Handset Alliance. Google, after purchasing the original developer of Android, released the operating system under the Open Handset Alliance.

Q: Is Android capable of running any Linux software?

A: Not necessarily. While I am sure that there will be ways to get around most any open source system, applications need to be compiled using the Android SDK to run on Android. The main reason for this is that Android applications execute files in a specific format; this will be discussed in later chapters.

Chapter 2

Downloading and Installing Eclipse

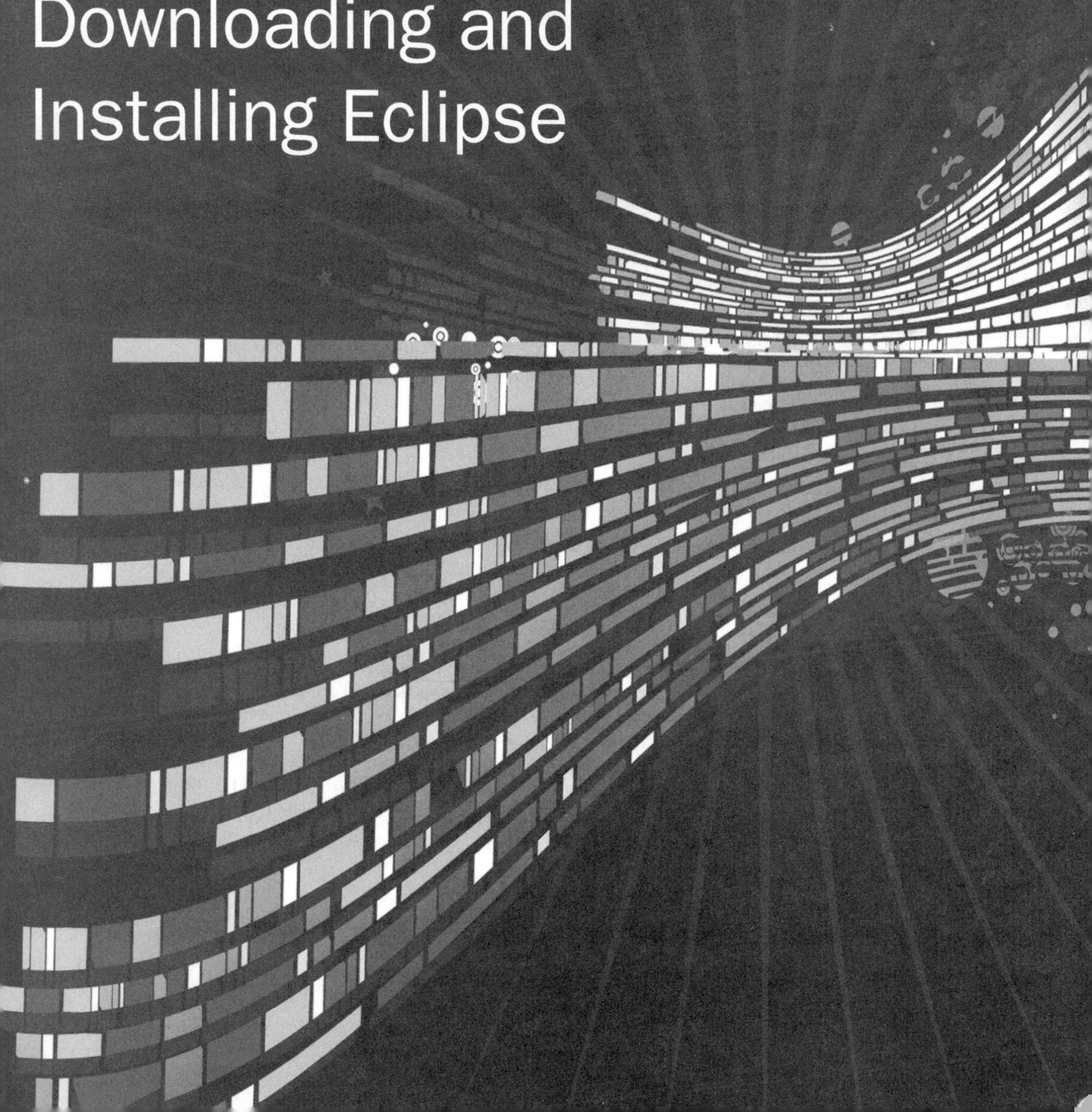

Key Skills & Concepts

- Selecting a development environment
- Downloading Eclipse
- Installing and configuring Eclipse

Android applications are developed in Java. Android itself is not a language, but rather an environment within which to run applications. As such, you can theoretically use any distribution or integrated development environment (IDE) you have at your disposal to begin your development. In fact, you can choose to use no IDE at all.

TIP

In later chapters of this book, I will give you an introduction to developing Android applications without the use of an IDE—or "in the command-line interface (CLI)." While I will not cover every example in the book using this technique, you will get the basics of how to develop in the CLI.

If you are more comfortable with one Java IDE over any other, such as JBuilder by Borland or the open source NetBeans, feel free to use it. With a moderate level of experience, you should still be able to follow along with the majority of the examples in this book. However, the Open Handset Alliance and Google do endorse one Java IDE over any others: Eclipse.

NOTE

If you choose to follow the examples in this book without using Eclipse, you need to check your IDE's documentation for compiling and testing your Android apps. The examples in this book give instructions only for compiling and testing in Eclipse, using the Android plugin for Eclipse.

This chapter concisely outlines the steps for downloading and installing Eclipse and the required Java Runtime Environment (JRE). Too many times, installation guides and tutorials, in a quest to not shut out more technologically advanced readers, tend to skip simple steps such as this. I have found that texts that skip these smaller steps often overlook important items. For this reason, I am including all of the download and installation steps in this chapter.

Why Eclipse?

Why is Eclipse the recommended IDE for Android applications? There are a few reasons for this particular endorsement:

- In keeping with the Open Handset Alliance's theme of truly opening the mobile development market, Eclipse is one of the most fully featured, free, Java IDEs available. Eclipse is also very easy to use, with a minimal learning curve. This makes Eclipse a very attractive IDE for solid, open Java development.
- The Open Handset Alliance has released an Android plugin for Eclipse that allows you to create Android-specific projects, compile them, and use the Android Emulator to run and debug them. These tools and abilities will prove invaluable when you are creating your first Android apps. You can still create Android apps in other IDEs, but the Android plugin for Eclipse creates certain setup elements—such as files and compiler settings—for you. The help provided by the Android plugin for Eclipse saves you precious development time and greatly reduces the learning curve, which means you can spend more time creating incredible applications.

NOTE

Eclipse is also available for Mac and Linux. Having greater availability, on numerous operating systems, means that almost anyone can develop Android applications on any computer. However, the examples and screenshots in this book are given from the Microsoft Windows version of Eclipse. Keep this in mind if you are using Eclipse in a non-Microsoft environment; your interface may look slightly different from the screenshots, but the overall functionality should not change. If there is a major change in operation of Eclipse under Linux, I will include an example of that change. I will provide several examples from within a Linux environment. The majority of these examples will be from the Linux/Android command-line environment.

Downloading and Installing the JRE

Before you begin downloading and installing Eclipse, you have to make sure you have the Java Runtime Environment (JRE) downloaded and installed on your machine. Because Eclipse as an application was written in Java, it requires the JRE to run. If the JRE is not installed or is not detected, you will see the following error if you try to open the Eclipse environment:

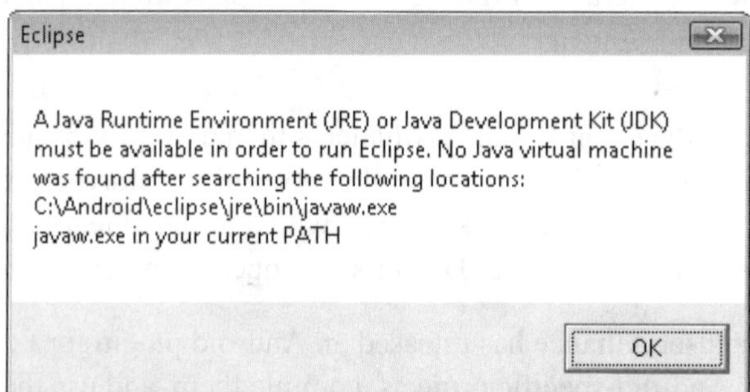

If you are an existing Java developer and already have Java installed on your computer, you will still want to follow along here, just to be sure you have the correct version of the JRE installed.

NOTE

Most people who have used the Web, or applications that are web-based, have the JRE installed. The JRE allows you to run Java-based applications, but it does not allow you to create them. To create Java applications, you need to download and install the Java Development Kit (JDK), which contains all the tools and libraries needed to create Java applications. If you are not familiar with Java, keep these differences in mind. For the examples in this book, I will be downloading the JDK, because it also includes the JRE. Although you don't need the JDK to run Eclipse, you can use it for other development later in the book.

Navigate to the Sun Developer Network (SDN) Downloads page at http://developers.sun.com/downloads/, as shown in the following illustration. Normally you only need the JRE to run Eclipse, but for purposes of this book you should download the full JDK, which includes the JRE. The reason for downloading the full JDK is that later in the book I will also give some examples of how to create Android applications outside Eclipse, using just the JDK tools. If you want to follow along with these tutorials, you will need the full JDK.

From the SDN Downloads page, navigate to the download section for the proper JDK. Select and initiate the download, as shown in the following illustration:

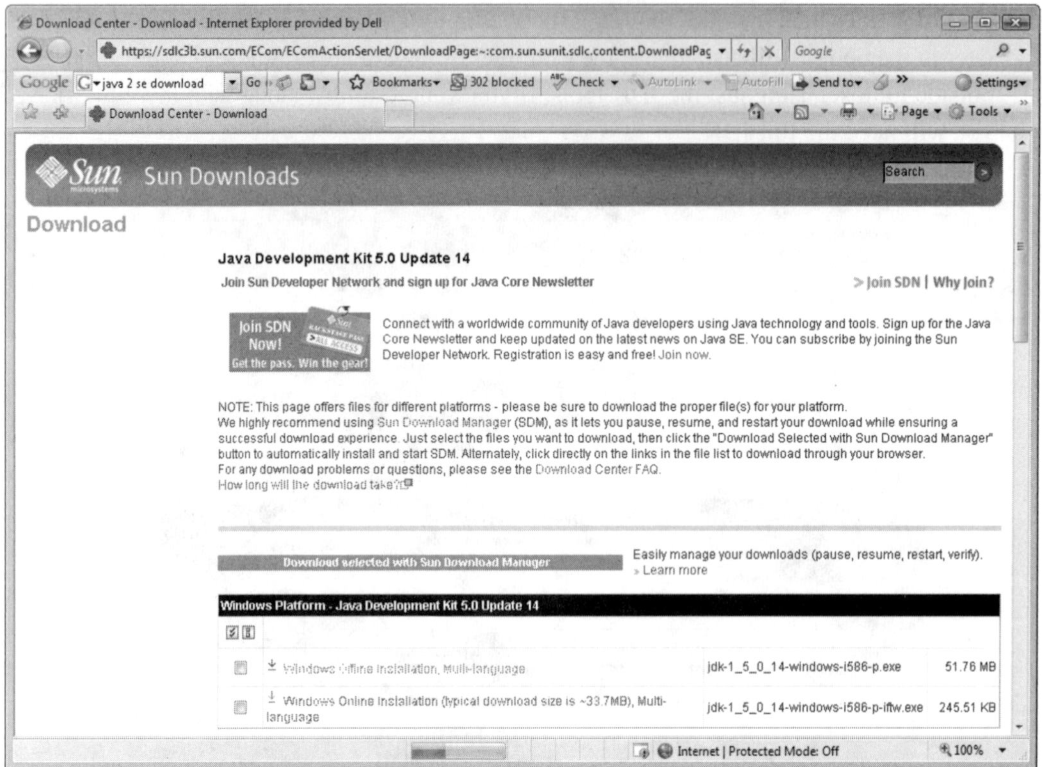

For the examples in this book, I chose to go with the Java 5 JDK update 14 because it is explicitly defined in the Eclipse documentation as the "supported" version of Java. To download the Java 5 JDK, select the platform for which you want to download. You should be able to follow along just as easily if you choose to download the Java 6 JDK. However, if you do want to download the older JDK 5, you need to click the Previous Releases link, as shown next:

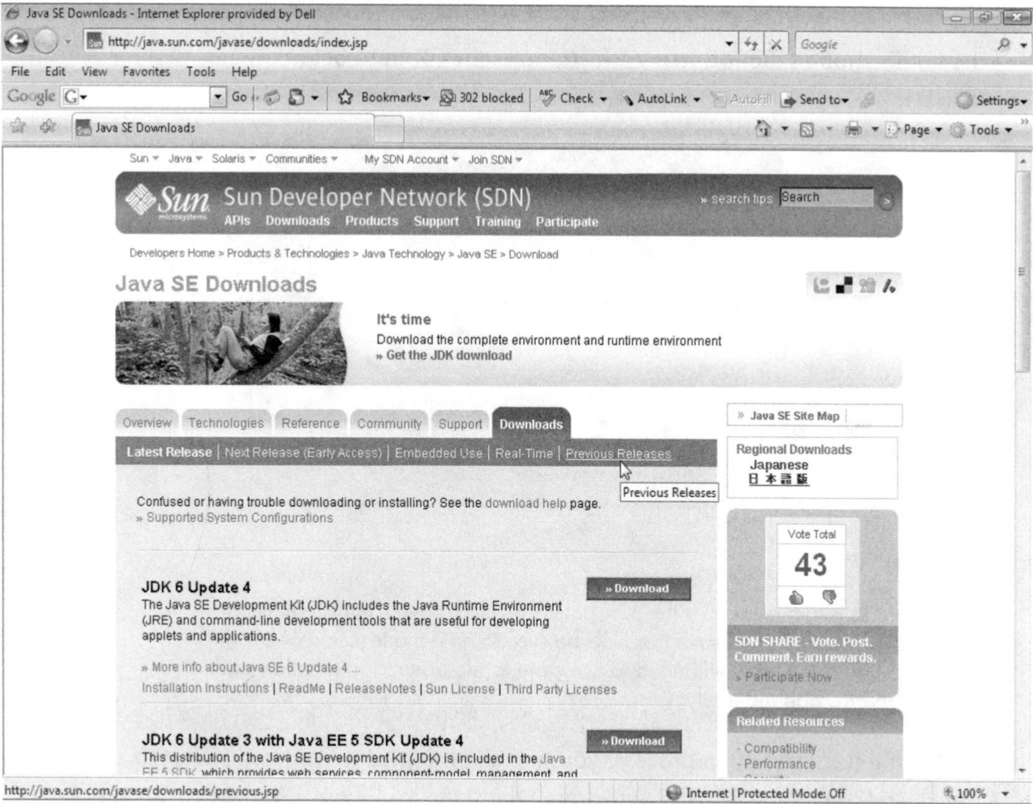

NOTE

You must agree to and accept the Sun licensing agreement on this page before you can initiate your download.

On the Java SE Previous Releases Downloads page, click the J2SE 5.0 Downloads link, and then click the Download button for JDK 5.0 Update *x*, where *x* is the latest update number (14 at the time of this writing but likely different by the time you read this).

If you are downloading to a Microsoft Windows environment, when you see the notification in the following illustration, click Run to begin the installation of the JDK.

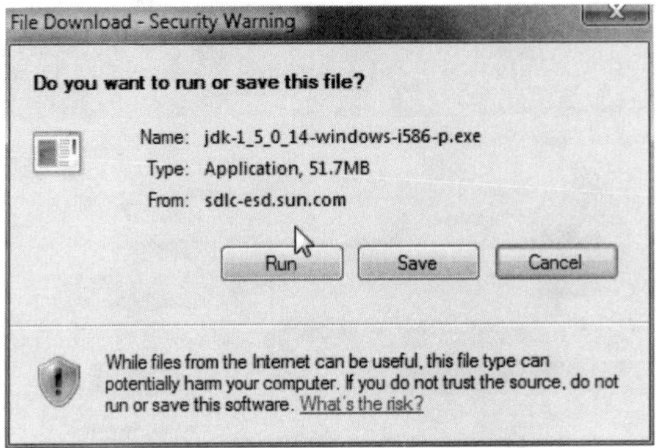

CAUTION

If you want to retain a copy of the JDK package, click Save rather than Run. However, if you choose to save the JDK, be sure to note the location. After the download completes, you will need to navigate to the download location and execute the package manually.

During the installation process, you will be prompted to read and accept the License Agreement, shown next. After agreeing to the standard License Agreement and clicking Next, you will be able to select your custom setup options.

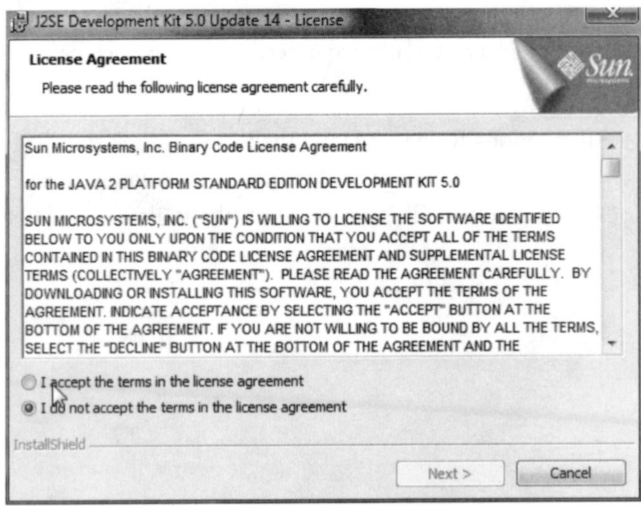

There is very little you need to change here, unless you are a more seasoned Java veteran and have particular options that you want to choose, in which case you should feel free to change the selections as you see fit. The following illustration shows the Custom Setup screen for the Java JDK.

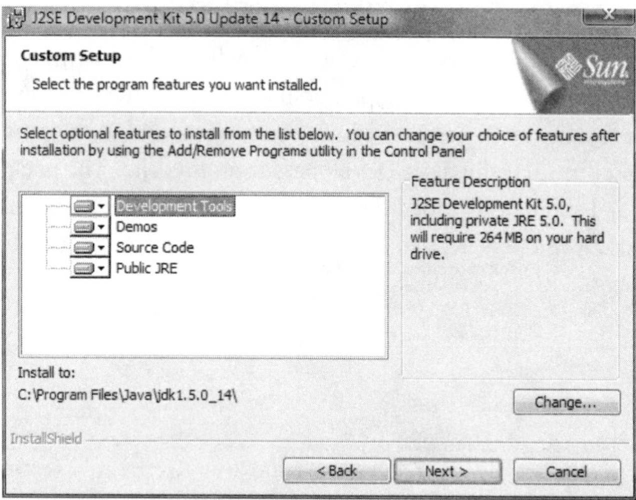

To keep the process simple, and fairly standardized, you should accept the suggested packages—by default everything is selected—and continue the installation by clicking Next. Once again, if you are comfortable with making specific changes, feel free to do so. However, if you have trouble in later chapters, you will want to modify your installation options. When the Installation Completed page appears, shown in the following illustration, click Finish and your installation should be completed.

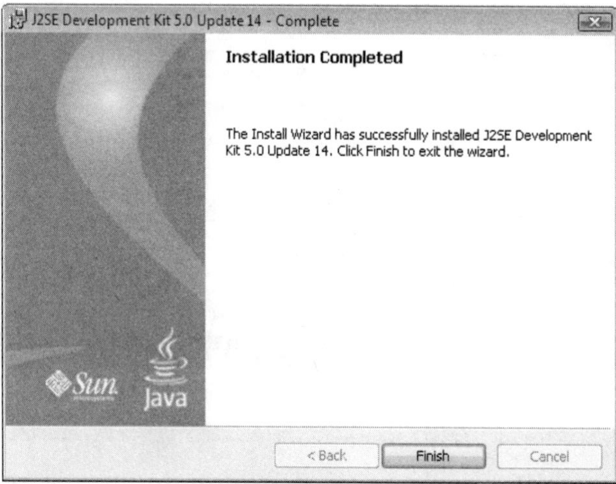

Once you complete the Java JDK installation—and by default the JRE installation—you can begin to install Eclipse.

Downloading and Installing Eclipse

Navigate to the Eclipse Downloads page at www.eclipse.org/downloads, shown in the following illustration. As the opening paragraph states, the JRE is required (Java 5 JRE recommended) to develop in Eclipse, which you took care of in the previous section. Download the Eclipse IDE for Java Developers from this site. The package is relatively small (79MB) and should download fairly quickly. Be sure not to download the Eclipse IDE for Java EE Developers, as this is a slightly different product and I will not be covering its usage.

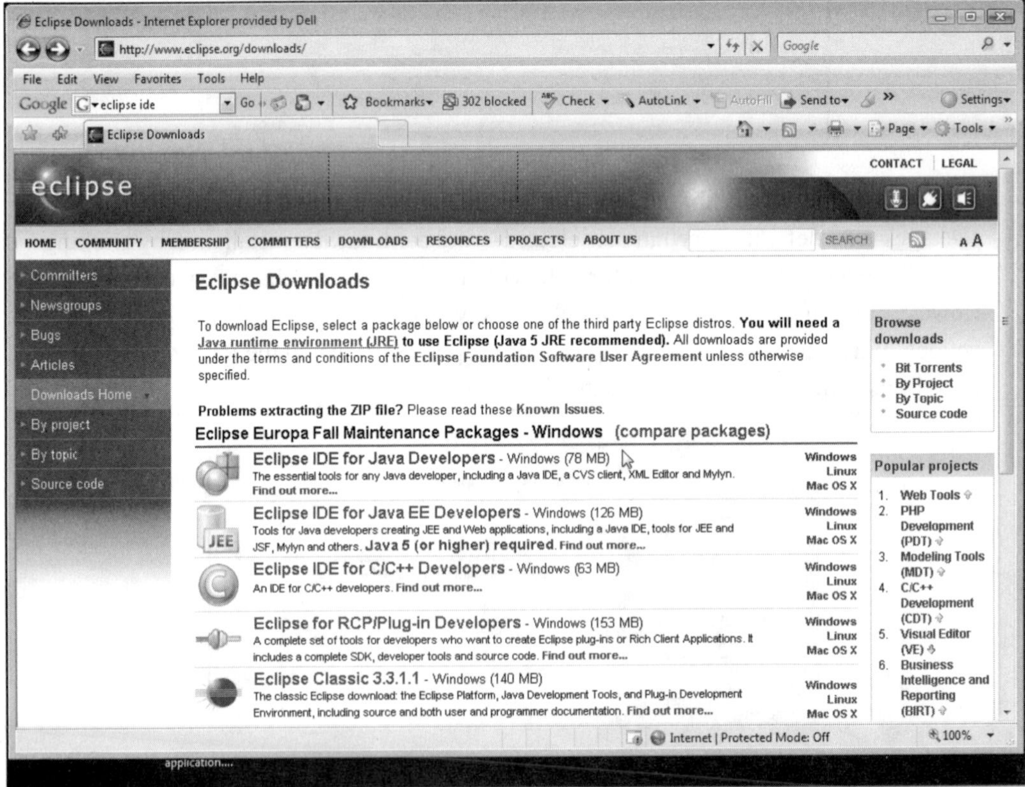

After you have downloaded Eclipse, it is time to install it. Navigate to the location where you downloaded the Eclipse package. As of the writing of this book, the latest Eclipse package file for Microsoft Windows is eclipse-java-europa-fall2-win32.zip. Expand the package and run the eclipse.exe file. Eclipse installs to your User directory by default (under Microsoft Windows), but you may want to install it to your Program Files directory. This will keep your applications in order and still allow you to set a different location for your workspaces. The following illustration shows the Eclipse title screen that appears upon startup.

NOTE

If you do not see the splash screen shown in the illustration, try rebooting your machine. If rebooting does not help, download and install the Java 5 JRE only.

Once the Eclipse installation commences, you will be prompted to create a default workspace, or folder. Just as in most development environments, projects are created in, and saved to, a workspace. The default path for the workspace is your User directory, as shown in the illustration that follows. To select a different location, click Browse and navigate to it.

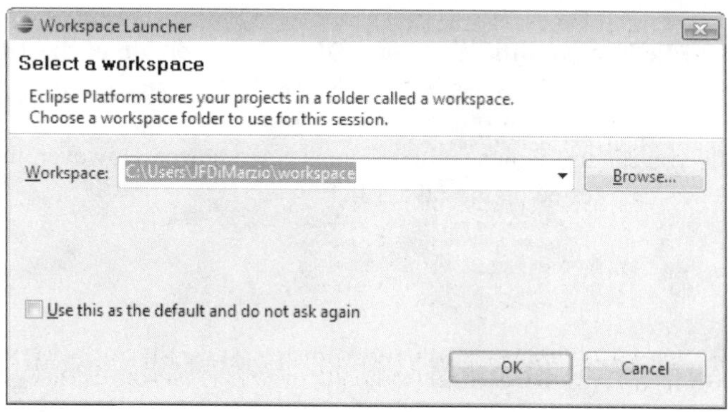

I recommend that you also check the check box that defaults all of your projects to the specified workspace. By checking this box, you will have one less thing to worry about when creating new projects and you will always know in what directory structure to find your source files. In this book, sometimes you will be navigating to the project files to work on them outside of the Android development environment, so knowing exactly where they are will be helpful.

After you select a location for your workspace, click OK.

At this point, your development environment is downloaded and installed. While the installation of Eclipse seemed deceivingly quick, you still need to do some configuration work before you can create your first Android project. Much of the configuration work that you need to do centers on the Android SDK and the Android plugin for Eclipse.

Next you need to download and install the Android SDK, download and install the Android plugin for Eclipse, and configure the Eclipse settings. By the end of Chapter 3 you will have a fully configured development environment within which you can begin to create your applications. You will then explore the Android SDK and begin creating your first Hello World! application in Chapter 5.

Ask the Expert

Q: Eclipse is used to develop applications in Java, but can Android run applications written in any other languages?

A: As of the writing of this book, there were no other SDKs or emulators available to allow Android development in any language other than Java.

Q: Can you use Eclipse (and the Android SDK) with a version of the JRE other than version 5?

A: Technically you can use Eclipse with versions 5 and newer. However, the latest version of Eclipse was only tested on the Java 5 JRE.

Chapter 3

Downloading and Installing the Android SDK

Key Skills & Concepts

- Downloading the Android SDK
- Using the Update feature of Eclipse
- Downloading, installing, and configuring the Android plugin for Eclipse
- Checking the PATH statement

In the previous chapter, you downloaded and installed your primary development environment, Eclipse. Now that your initial development environment is established, using Eclipse as your Java IDE, you can use it to develop Java applications, but you have one more step before you can begin creating mobile phone applications. You must configure it in a way that will facilitate Android development.

Because Eclipse is a Java development environment, you can create and edit Java projects with great ease. However, given that you have no libraries yet for understanding how Android applications should behave, you cannot develop anything that will run on an Android-based device. To begin creating Android projects, you need to download and install the Android SDK. You must then download the related Android plugin for Eclipse to utilize the SDK within the Eclipse IDE. With these pieces in place, you can begin your development.

If you have any development experience, you are most likely familiar with the process of using an SDK. Desktop application developers, regardless of the platform they are developing on, use SDKs to create applications that will run on the desired system they are developing on. The Android SDK is no different from any other SDK in that it contains all the Java code libraries needed to create applications that run specifically on the Android platform. The SDK also includes help files, documentation, an Android Emulator, and a host of other development and debugging tools.

NOTE

Chapter 4 covers most of the functionality of the Android SDK in depth.

To begin, you are going to download the Android SDK from the Google Android development site, located at http://code.google.com/android. The Google Android development home page contains a host of valuable tools and documents about developing for the Android platform, including links to the Android developer forum (or "community"). Figure 3-1 shows the home page for Google Android development.

TIP

If you ever encounter a problem while you are developing an Android application, the first place you should look for an answer is the Android developers' forum at http://code.google.com/android/groups.html. There are discussion groups for beginners, developers, and "hackers," and a general-issue discussion group. Given that Android is such a new platform, the Android developers' forum is one of the few places to find comprehensive, reliable information about developing for the product.

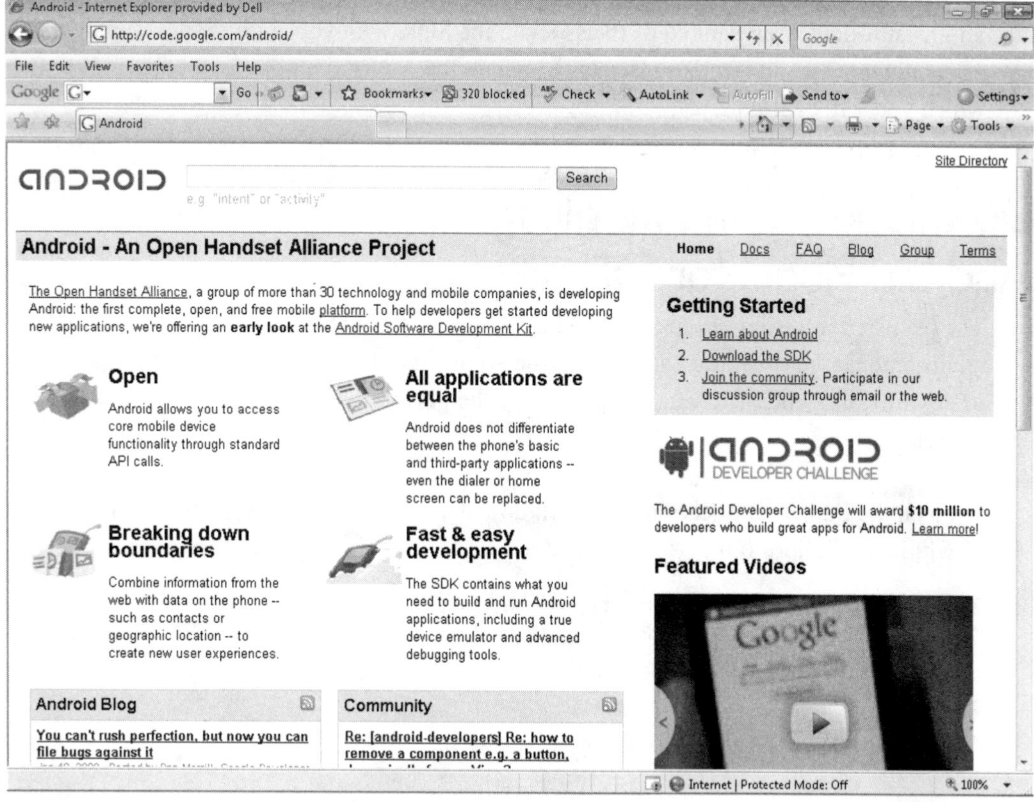

Figure 3-1 The Google Android development home page.

Downloading the Android SDK

The Android SDK is easily accessible from the http://code.google.com/android page. From the development home page, click the Download the SDK link under Getting Started. After you agree to the terms of the Android SDK License Agreement, you will see the Download the Android SDK page. The Android SDK is downloaded in a 79MB (for Windows) package and it should download fairly quickly. Click the package name for your operating system to begin the download.

NOTE

Download sizes for other operating systems may vary.

There is no "setup" or installation process to speak of for the Android SDK; rather, you must follow a series of steps to associate the SDK with your Eclipse development environment. The first of these steps is to obtain the Android plugin for Eclipse, after which you will configure it.

Downloading and Installing the Android Plugin for Eclipse

The first step in setting up the Android SDK within the Eclipse development environment is to download and install the Android plugin for Eclipse. Both tasks of downloading and installing the plugin can be performed at the same time, and are relatively easy to do:

1. Open the Eclipse application. You will download the Android plugin for Eclipse from within the Eclipse IDE.
2. Choose Help | Software Updates | Find and Install.

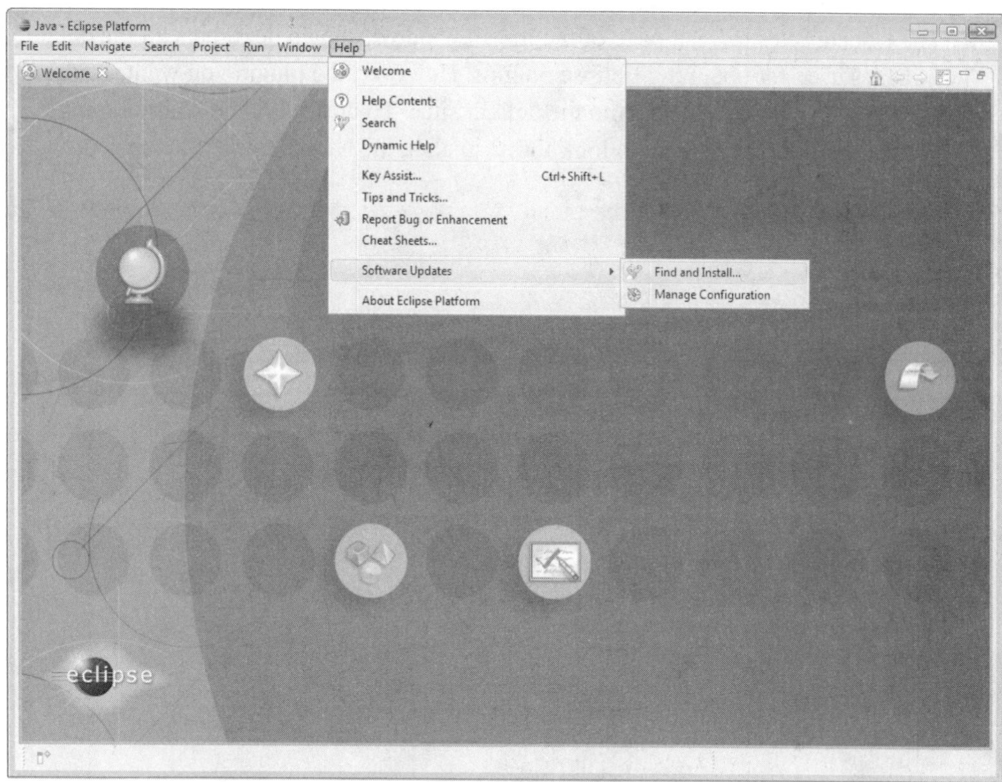

3. In the Install/Update window, which allows you to begin the process of downloading and installing any of the plugins that are available to you for Eclipse, click the Search for New Features to Install radio button and then click Next.

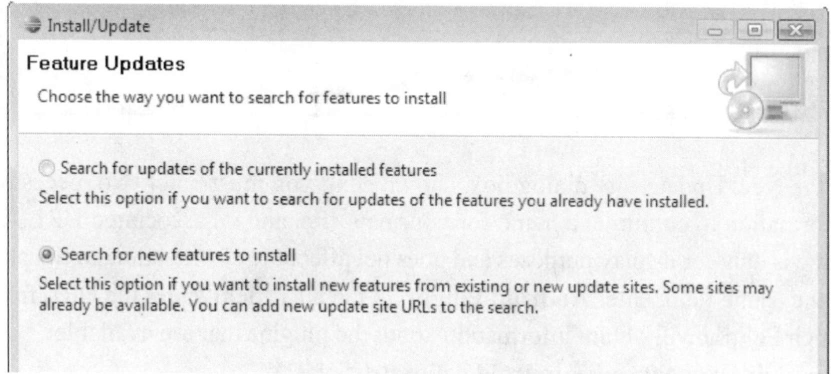

4. The Update Sites to Visit page of the Install window, shown next, lists all the default websites used for obtaining Eclipse plugins. However, the plugin you want, Android for Eclipse, is not available from the default sites. To download the Android plugin, you must tell Eclipse where to look for it, so click the New Remote Site button.

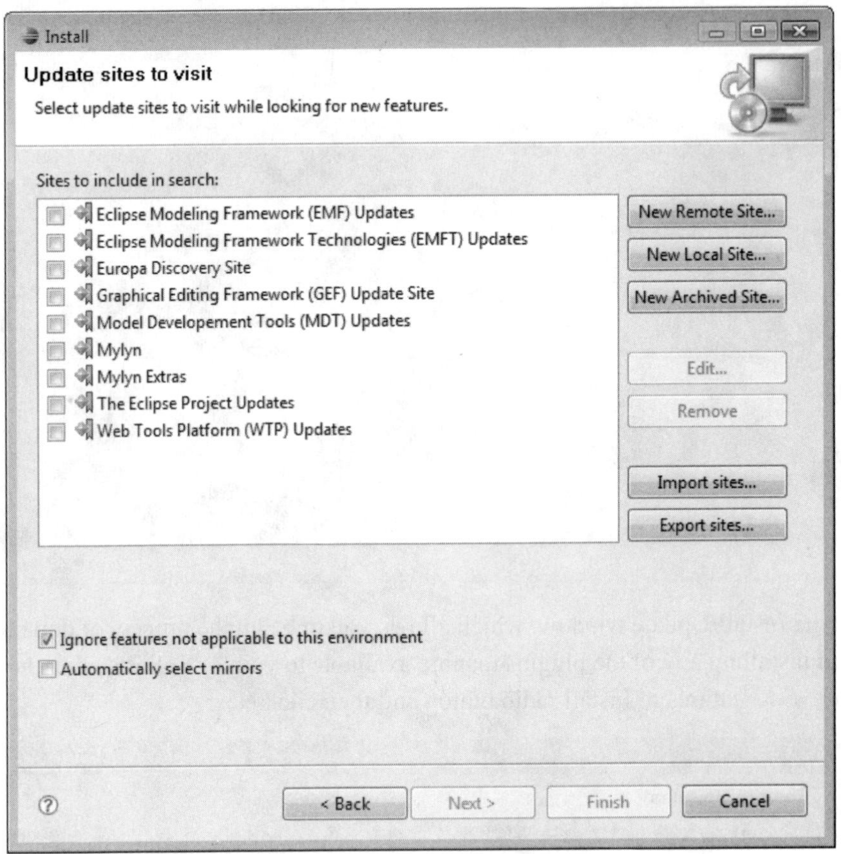

5. In the New Update Site dialog box, shown next, you must enter two pieces of information to continue: a name for your new site, and its associated URL. The name is only for display purposes and does not affect the downloading of the plugin. In the Name field, enter **Android Plugin**. In the URL field, enter the URL from which Eclipse will obtain information about the plugins that are available: **https://dl-ssl.google.com/android/eclipse/**. Click OK.

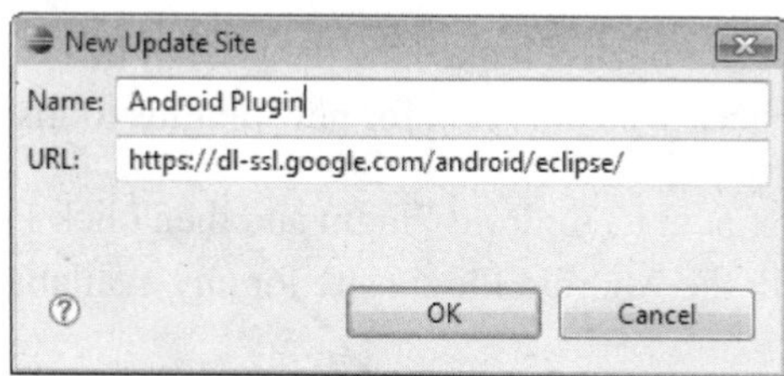

NOTE

The name for your site can be anything you want, as long as it will help you identify what the link is. Feel free to use something other than Android Plugin.

6. A new site named Android Plugin should now be in your list of available sites:

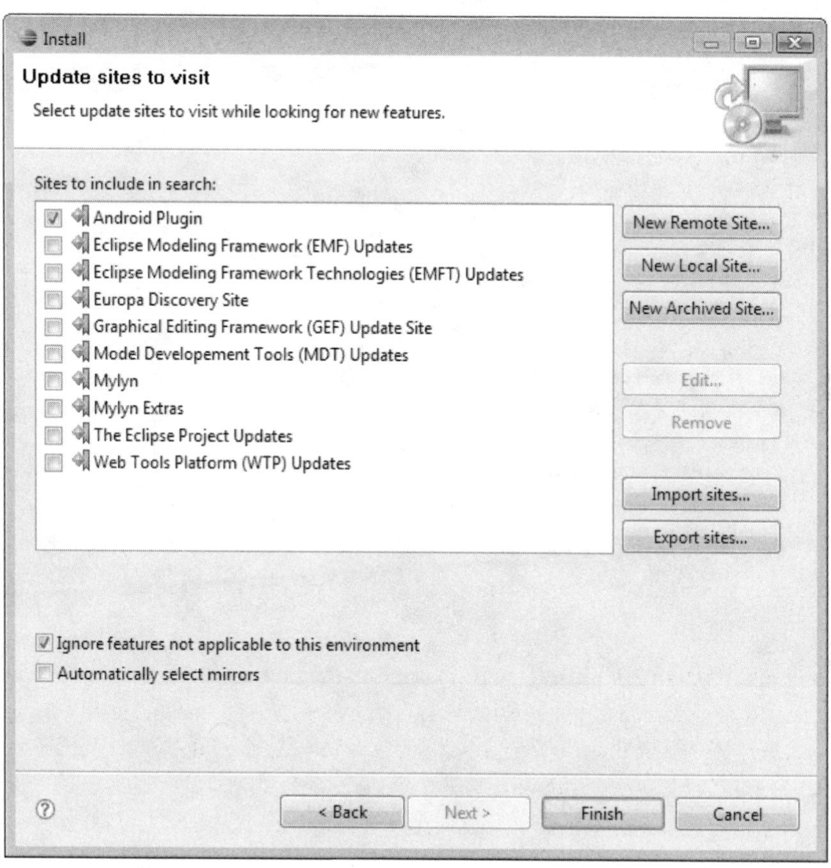

At this point Eclipse has not yet looked for the plugin; this is just a list of paths that you can tell Eclipse to check when looking for new plugins to install.

7. Check the check box next to Android Plugin and then click Finish. Eclipse searches the URL associated with the Android Plugin site for any available plugins.

8. On the Search Results page of the Updates window, select the Android Plugin and then click Finish.

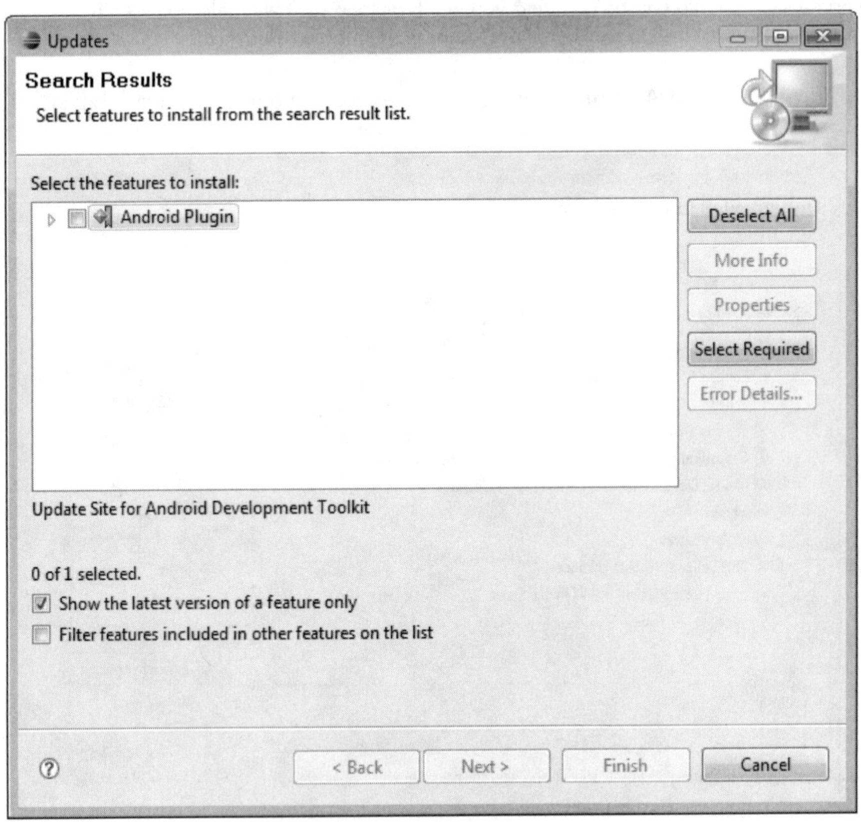

9. On the Feature License page of the Install window, shown next, accept the licensing agreement for the Android Development Tools and click Next.

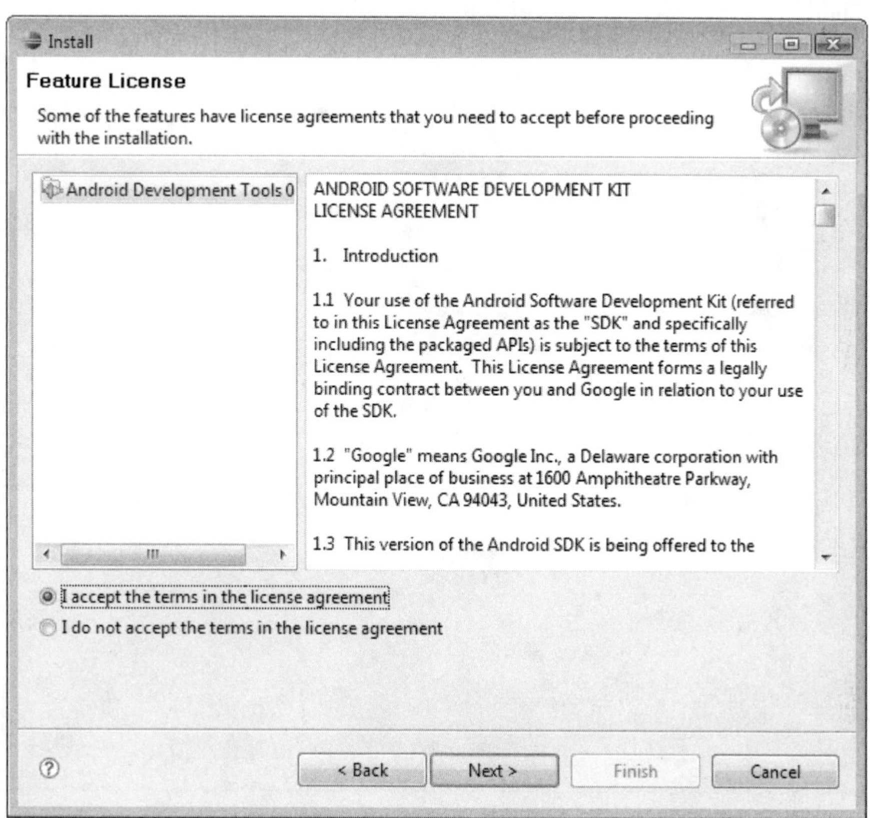

NOTE

Keep in mind that all Eclipse plugins are installed to the %installpath%/eclipse/plugins directory. This information will help you if you need to locate the files that make up the Android plugin.

10. Eclipse downloads the Android plugin. At the time of this writing, the plugin version is 0.4.0.200802081635. On the final plugin installation page, Feature Verification, click Install All to complete the installation of the Android plugin.

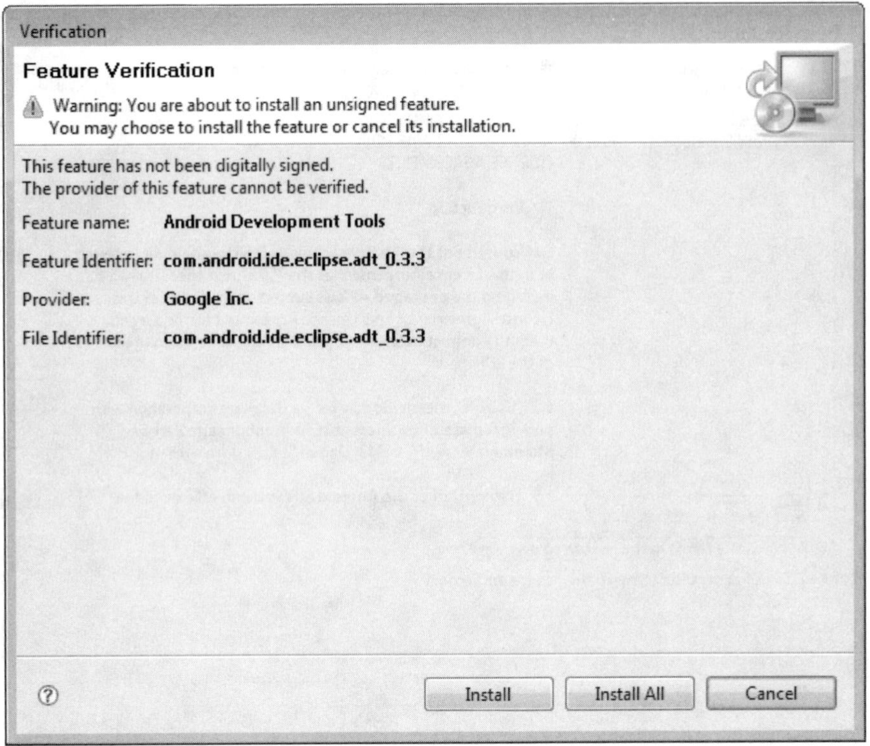

With the Android plugin installed, the last step you have to perform is to configure the plugin.

Configuring the Android Plugin for Eclipse

After installing the Android plugin for Eclipse, Eclipse should have prompted you to restart the application. If it did not prompt you, restart Eclipse now. Restarting Eclipse will ensure that the program has a chance to reinitialize with the plugin installed. It is important to make sure configuration steps like this are followed in order to reduce the chance of misconfigurations.

The Android plugin for Eclipse is configured from the Preferences window of Eclipse. Proceed as follows:

1. From the main Eclipse window, choose Window | Preferences.
2. In the Preferences window, shown next, select Android in the menu on the left. On the right side of the window, click Browse, find the location of the Android SDK on your hard drive, and enter it in the SDK Location field. Eclipse needs this information to be able to access all the tools that are supplied with Android, such as the emulator.

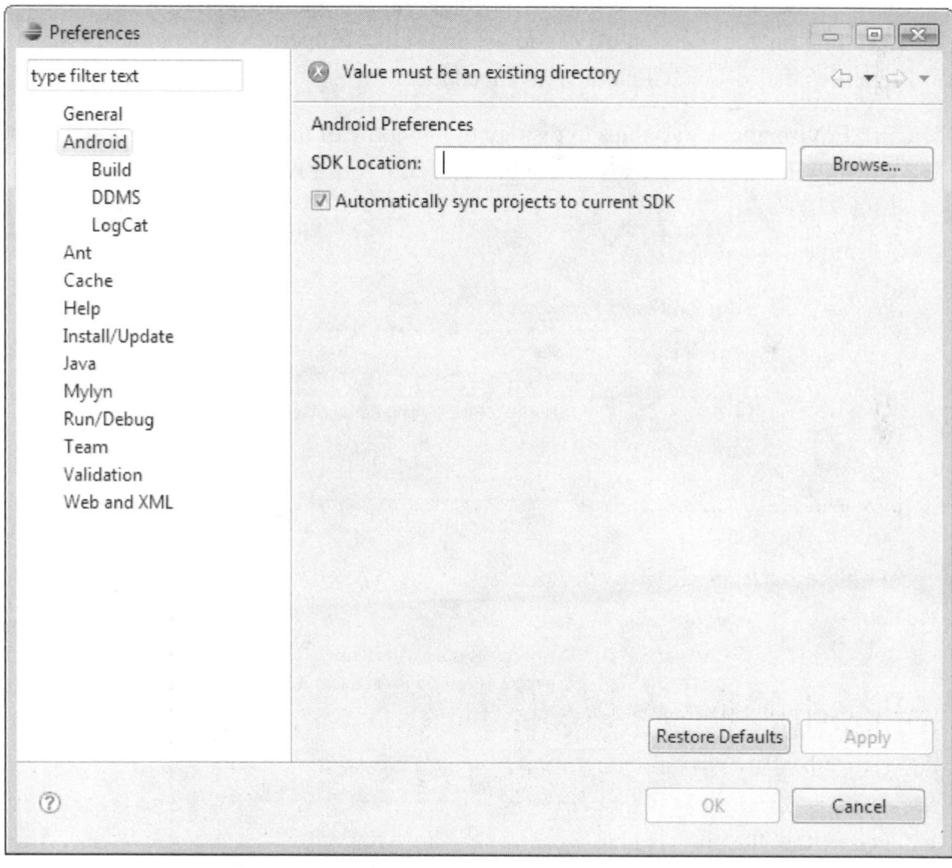

3. Check the Automatically Sync Projects to Current SDK check box and then click Apply.

NOTE

The Android plugin for Windows is shipped in a zip file that contains a directory with a very long directory name: android-sdk_m5-rc14-win32. It may help you in future chapters, especially when command-line programming, to rename this directory to something more manageable. You may also want to extract the SDK to the Program Files directory.

4. The final step in setting up the Android SDK is to put it into your PATH statement. If you are using a Microsoft Windows machine, right-click Computer (or My Computer, depending on your version of Windows) and select Properties to open the System Properties dialog box. Click the Advanced tab.

5. Click Environment Variables to display the window of the same name, shown in the following illustration. This is where you can edit your PATH statement.

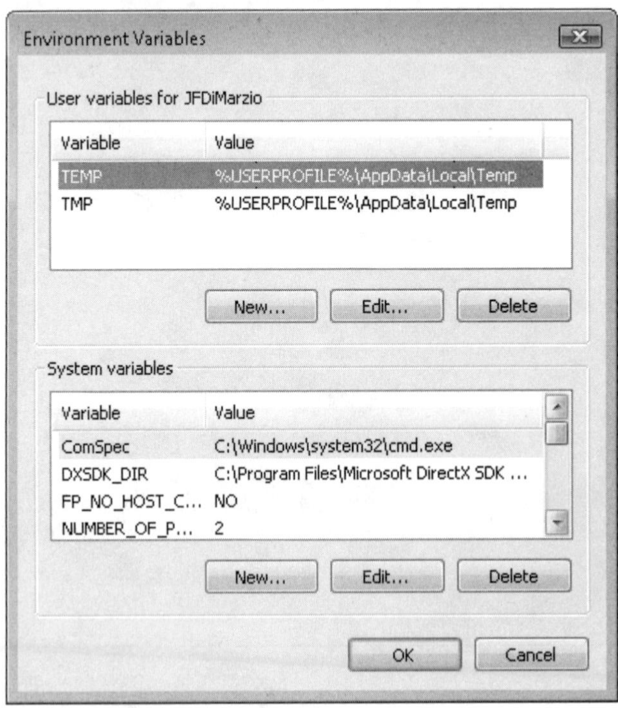

6. Under System Variables, find the variable PATH and double-click it.
7. In the Edit System Variable dialog box, shown next, add the location of your Android SDK—separated from the existing paths by a semicolon—and click OK to apply your changes, and click OK again in the Environment Variables window.

The Android SDK, Eclipse, and the Android plugin for Eclipse are now fully configured and ready for development. In the next chapter, you will explore the Android SDK, and learn about its features. The Android SDK contains many tools to help you develop full-featured cell phone applications, and the next chapter provides a good overview.

Ask the Expert

Q: Is the Android SDK available for any languages other than Java?

A: No. Android applications can be developed only in Java.

Q: Will there be updates to the Android SDK?

A: Yes! Even during the writing of this book, an SDK update was released that addresses many issues within the platform. I suggest checking the development page often for the latest updates.

(continued)

Q: If an update is released, how do I upgrade my SDK?

A: Upgrading the SDK can be very tricky. When a new SDK is released, chances are a new plugin is also released. During the writing of this book, both a new SDK and a new plugin were released. I attempted to use the "provided" upgrade tools to change versions. However, this proved fruitless and left me with two conflicting versions, neither of which worked correctly. I eventually had to uninstall both versions and reinstall only the latest version. The newest SDK then worked correctly. I suggest that anyone faced with the possibility of upgrading from one version of an SDK/plugin combo to another use this same process: simply uninstall the older version, and install the newer one, rather than upgrading.

Chapter 4

Exploring the Android SDK

Key Skills & Concepts

- Using the Android SDK documentation
- Using the Android SDK tools
- Using the sample applications
- Learning the life cycle of an Android application

Now that you have your development environment established, you are ready to explore the Android SDK, which contains multiple files and tools specifically intended to help you design and develop applications that run on the Android platform. These tools are very well designed and can help you make some incredible applications. You really need to be familiar with the Android SDK and its tools before you begin programming.

The Android SDK also contains libraries for tying your applications into core Android features such as those associated with cell phone functions (making and receiving calls), GPS functionality, and text messaging. These libraries make up the core of the SDK and will be the ones that you use most often, so take the time to learn all about these core libraries.

This chapter covers all of the important items contained within the Android SDK. By the end of the chapter, after familiarizing yourself with the contents of the Android SDK, you will be comfortable enough to begin writing applications. However, as with any subject, before you can dive into the practice of the discipline, you must familiarize yourself with its contents and instructions.

CAUTION

I am not going to go over every minute detail of the Android SDK; Google does a very good job of that within its documentation. To avoid the risk of spending too much time *discussing* how things work instead of *showing* you how they work, I have tried to keep this discussion as brief as possible. I cover only the most important topics and items, leaving you free to explore the rest in more depth yourself, at your own pace.

What Is in the Android SDK?

The Android SDK is downloaded in a simple zipped package (as described in Chapter 3). The bulk of the Android SDK, in number of files, consists of documentation, with programming APIs, tools, and samples comprising the rest. This section provides a closer look at exactly what is included in the Android SDK.

TIP

Chapter 3 suggested that you extract the Android SDK to the Program Files folder, so that it would be easier to track. If you are having trouble finding the SDK because you used the default extraction setting, it should be in the following folder: /%downloadfolder%/android-sdk_m5-rc14_windows/android-sdk_m5-rc14_windows.

Navigate to the folder where you unpacked the Android SDK so that you can begin to explore the folder structure within. While there are a few files in the root folder, like android.jar (a compiled Java application containing the core SDK libraries and APIs) and some release notes, the remainder of the Android SDK is divided into three main folders:

- **Docs** Contains all of the accompanying Android documentation

NOTE

Much of the documentation found in the Docs folder can also be found on the http://code.google.com/android Android development site.

- **Samples** Contains six sample applications that you can compile and test from within Eclipse
- **Tools** Contains all of the development, compilation, and debugging tools that you need throughout the development process of an Android application

The following sections discuss in a bit more detail what is included in each of the SDK folders. Each API demo is compiled and run to illustrate the capabilities of Android. Many of the tools are discussed and demonstrated in later chapters as you learn how to create and compile applications using the command-line options of Microsoft Windows and Linux.

Android Documentation

The Android documentation is located in the Docs folder within the Android SDK at ../%sdk folder%/DOCS. The documentation that is supplied with the SDK includes steps on downloading and installing the SDK, "Getting Started" quick steps for developing applications, and package definitions. The documentation is in HTML format and can be accessed though the documentation.html file in the root of the SDK folder. The following illustration depicts the main page of the Android SDK documentation.

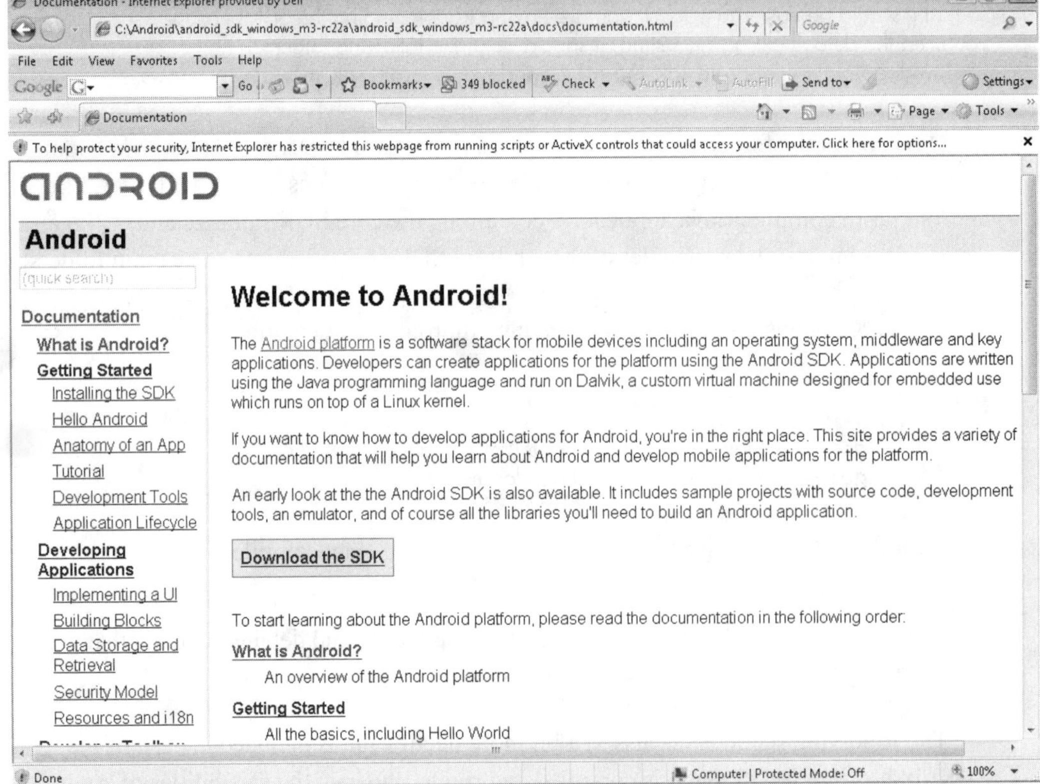

You can navigate to all of the documentation that is included in the Android SDK by using the links within documentation.html.

CAUTION

As you are navigating the Android SDK, you may think some pages are mislinked or missing, because the right side of the screen may be blank when you click some links. However, if you scroll down you will see that the pages are just misaligned.

In working with the Android SDK, I have found that there are sections of the documentation that I refer to more than others. For me, the most valuable segments of the Android SDK documentation are as follows (as they appear in the navigation bar):

- **Reference Information**
  - Class Index
  - List of Permissions
  - List of Resource Types
- **FAQs**
  - Troubleshooting

The Troubleshooting subsection of the documentation will be especially helpful as you are starting out. As you progress through the book and begin to develop your own Android applications, you will find that the Reference Information section of the documentation is more helpful. For example, while it would have little to no use to you now, the List of Permissions subsection will be very helpful to you when you progress to the section of the book that deals with creating more complex applications. Take some time to familiarize yourself with the Android SDK documentation and the hundreds of documents that have been provided for you.

Android Samples

The Samples folder, ../%sdk folder%/SAMPLES, contains six sample applications that demonstrate a good cross-section of Android functionality:

- API Demos
- Hello, Activity!
- Lunar Lander

- Note Pad
- Skeleton App
- Snake

These sample applications are provided by Google to give you a quick idea of how to develop an Android application. Each sample application demonstrates a different piece of Android's functionality. You can open and run these applications from within Eclipse. Following is a brief description of each.

API Demos

The API Demos application is a host application that demonstrates multiple API functions in a single Activity.

TIP

An *Activity* is an Android application. Activities are covered in more depth in the following chapters.

The API Demos application, as shown in the following illustration, contains multiple, smaller, examples of different Android functions:

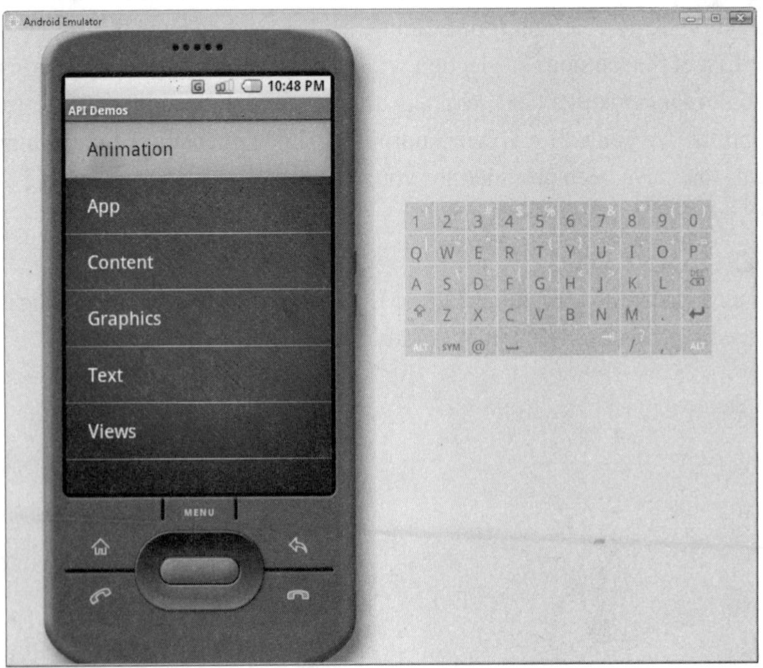

Some of the applications included in the API Demos sample include 3-D image transitions, list and progress dialog boxes, and a finger-painting demo.

Try This Run the API Demos Sample Application

Using Eclipse, load the API Demos application as a New Android Project. To do this, select File | New | Project from the Eclipse menu bar; a New Android Project wizard opens. Do not worry about the options in this wizard for now. Simply select Create Project From Existing Source and browse to the folder with the API Demo application in it. When the project is loaded, choose Run to see it execute in the Android Emulator.

Navigate your way through the more than 40 different applications. Use each application to become familiar with the terminology and function of each API tool it demonstrates.

Hello, Activity!

The Hello, Activity! application, shown in the following illustration, is a simple Hello World!–style application. Though simple in its design, Hello, Activity! does a good job of showing off the abilities of the platform. You will create your own Hello World!–style applications in the next chapter.

Lunar Lander

Lunar Lander, shown next, is a small game that plays on the Android Emulator. Lunar Lander shows how a simple 2-D game works on Android. The controls are fairly simple, and the game is not very complex. However, given these drawbacks, it is a great starter for game development.

Lunar Lander implements a simple control scheme (Up, Down, Left, and Right). The game also displays relatively fluid graphics and looks impressive given the platform. Complex game theories such as collision detection are used in a simple way. Although this book does not cover programming games for the Android platform, if you are interested in doing so, you may want to look at Lunar Lander for some tips.

Note Pad

Note Pad, as shown in the illustration that follows, allows you to open, create, and edit small notes. Note Pad is not a full-featured word editor, so do not expect it to be something to rival Word for Windows Mobile. However, it does a good job as a demonstration tool to show what is possible with a relatively small amount of code.

Skeleton App

Skeleton App, shown next, is an application shell. This is more of a base application that demonstrates a couple of different application features, such as fonts, buttons, images, and forms. If you are going to run Skeleton App by itself, you really are not going to get much out of it. You will be better served by referring to Skeleton App as a resource for how to implement specific items.

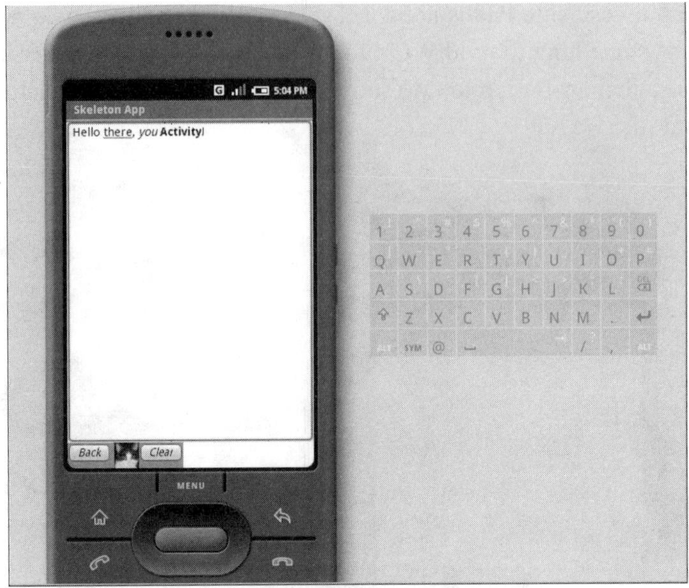

Snake

The final demo that is included with the Android SDK is Snake. This is a small, SNAFU-style game that is far more simplistic than Lunar Lander. This illustration shows what Snake looks like when run.

NOTE

If you navigate to the base folder of each of the sample applications, you will see a folder named src. This is the source code folder for the given sample application. You can use this to view, edit, and recompile the code for any of the applications. Take advantage of this source code to learn some tricks and tips about the Android platform.

Android Tools

The Android SDK supplies developers with a number of powerful and useful tools. Throughout this book, you will use only a handful of them directly. This section takes a quick look at just a few of these tools, which will be covered in much more depth in the following chapters, as you dive into command-line development.

NOTE

For more detailed information about the other tools included in the Android SDK, consult the Android doc files.

emulator.exe

Arguably one of the most important tools included in the Android SDK is emulator.exe. emulator.exe launches the Android Emulator. The Android Emulator is used to run your applications in a pseudo-Android environment. Given that, as of the writing of this book, there were no hardware devices yet released for the Android platform, emulator.exe is going to be your only means to test applications on a "native" platform.

You can run emulator.exe from the command line or execute it from within Eclipse. In this book, you'll usually let Eclipse launch the Android Emulator environment for you. However, in the interest of giving you all the information you need to program with the Android SDK outside of Eclipse, Chapter 6 covers command-line usage of emulator.exe when you create your Hello World! applications.

When using the Android Emulator to test your applications, you have two choices for navigating the user interface. First, the emulator comes with usable buttons, as shown in Figure 4-1. You can use these buttons to navigate Android and any applications that you develop for the platform.

TIP

The Power On/Off, Volume Up, and Volume Down buttons are slightly hidden to the sides of the virtual device. They identify themselves when you hover the mouse pointer over them.

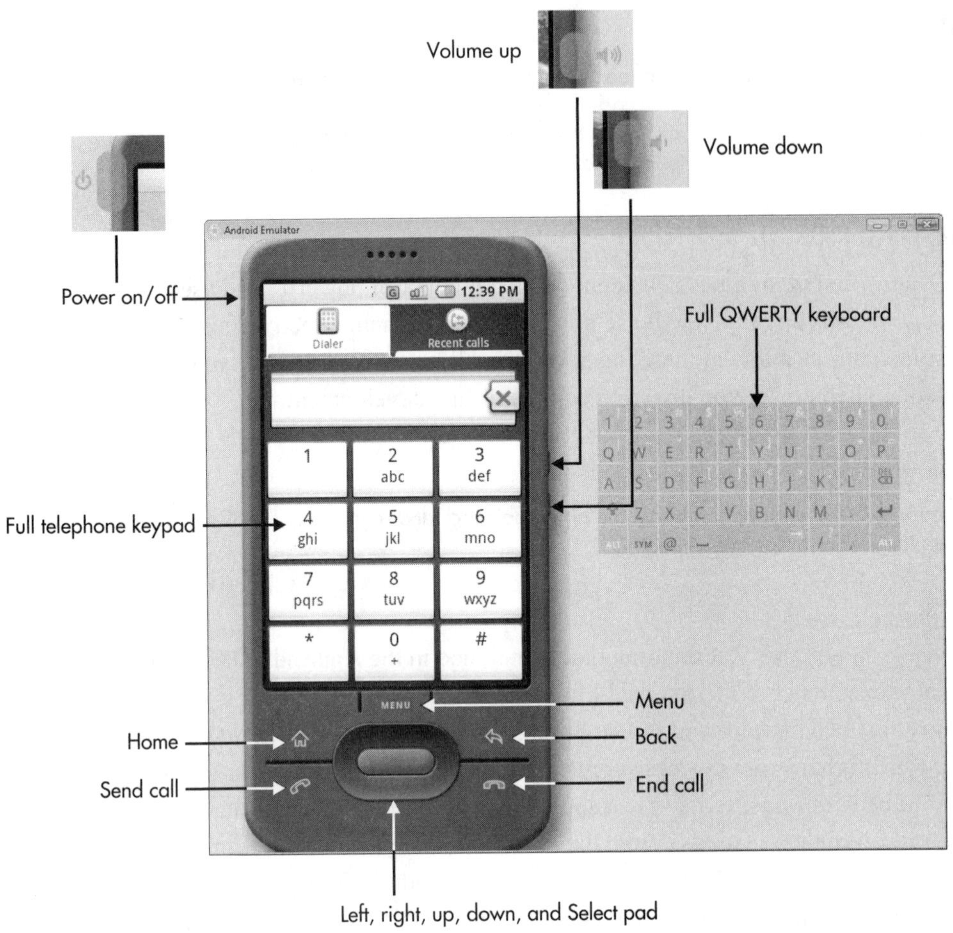

Figure 4-1 Navigating with the Android Emulator

Given that many higher-end phones now include a touch screen, the second input choice you have when using the emulator is a simulated touch screen. You use your mouse as a stylus. The objects on the emulator's screen can be interacted with using the mouse.

adb.exe

Another tool that will become very useful to you when you are using command-line programming is Android Debug Bridge or adb (adb.exe). This tool allows you to issue commands to the Emulator.exe tool. When you are working in a command-line environment, the adb tool allows you to do the following:

- Start and stop the server
- Install and uninstall applications
- Move files to and from the emulator

MKSDCARD.exe

MKSDCARD.exe is a very useful tool if you are testing an application that will need to read or write files to or from an SD Memory Card inserted into the mobile device. MKSDCARD.exe creates a small partition drive on your drive that will hold and retain the test files. The emulator will treat this partition like an SD Memory Card.

DX.exe

DX.exe is the compiler of the Android SDK. When run against your Java files, DX.exe will create files with .dex extensions—Dalvik executable format. These files are in the correct format to be understood by, and run on, an Android device.

NOTE

Android executable files are called Dalvik executable files as a reference to the Dalvik virtual machine that Android used to run all applications. The Dalvik virtual machine runs each application in its own thread with the same priority as core Android applications.

activityCreator(.bat or .pn)

activityCreator is a simple command-line tool that is used to set up a basic development environment. When run from the command line, activityCreator will set up the shell files needed to create a basic Android application. Having these shell files is especially useful if you are not using Eclipse. The Android plugin for Eclipse sets up these shell files for you by calling the activityCreator when you create a new project.

Depending on what type of environment you are running, you will see the activityCreator represented by a different type of script file. If you are in a Windows environment, this will be a .bat file; otherwise it will be a python (.pn) script. You simply execute the script, which in turn calls the actual activityCreator process with the correct parameters.

APIs

The API, or application programming interface, is the core of the Android SDK. An API is a collection of functions, methods, properties, classes, and libraries that is used by application developers to create programs that work on specific platforms. The Android API contains all the specific information that you need to create applications that can work on and interact with an Android-based application.

The Android SDK also contains two supplementary sets of APIs—the Google APIs and the Optional APIs. Subsequent chapters will focus much more on these APIs as you begin writing applications that utilize them. For now, take a quick look at what they include so that you are familiar with their uses.

Google APIs

The Google APIs are included in the Android SDK and contain the programming references that allow you to tie your applications into existing Google services. If you are writing an Android application and want to allow your user to access Google services through your application, you need to include the Google API.

Located in the android.jar file, the Google API is contained within the com.google.* package. There are quite a few packages that are included with the Google API. Some of the packages that are shipped in the Google API include those for graphics, portability, contacts, and calendar utilities. However, the packages devoted to Google Maps will be the primary focus in this book.

Using the com.google.android.maps package, which contains information for Google Maps, you can create applications that interact seamlessly with the already familiar interface of Google Maps. This one set of packages opens a whole world of useful applications just waiting to be created.

The Google API also contains a useful set of packages that allows you to take advantage of the newer Extensible Messaging and Presence Protocol (XMPP) developed by the Jabber open source community. Using XMPP, applications can quickly become aware of other clients' presence and availability for the purpose of messaging and communications. The API packages dealing with XMPP are very useful if you want to create a "chat"-style program that utilizes the phone messaging capabilities.

Optional APIs

The Android SDK includes a number of Optional APIs that cover functionality not covered by the standard Android APIs. These Optional APIs are considered optional because they deal with functionality that may or may not be present on a given handset device. That is, some devices created for the Android platform may include upgrades and features that others do not; the Optional APIs cover your programming options when trying to utilize these features in your Android applications.

One of these optional features (which you will use later in the book) is a cell-phone-based GPS. The Android LBS (Location-Based Services) API deals with the functionality needed to receive and utilize information from a device's GPS unit. (Combine the information in the Android LBS API with that in the Google Maps API, and you might have a very useful application that can automatically display a map of where you are located at any given point in time.)

Other Optional APIs include those for utilizing Bluetooth, Wi-Fi, playing MP3s, and accessing 3-D—OpenGL-enable hardware.

Application Life Cycle

If you have a decent amount of experience as an application developer, you are familiar with the concept of an application life cycle. An application life cycle consists of the steps that the application's processes must follow from execution to termination. Every application, regardless of the language it was written in, has a specific life cycle, and Android applications are no exception. This section examines the life cycle of an ASP application and compares that to an Android application's life cycle.

Standard ASP Application Life Cycle

The life cycle of a standard ASP application is similar enough to that of an Android application to make this a good comparison. ASP applications step through five distinct processes from launch to disposal. These steps are required to be implemented by all ASP applications, and really define what an ASP application is. The steps, in order, are

1. Application_Start
2. Event
3. HTTPApplication.Init
4. Disposal
5. Application_End

TIP

Some ASP reference materials consider Disposal and Application_End to be one step in the life cycle. However, the Disposal call can be intercepted before it is passed to Application_End. This can allow the application to perform specific functions before it is actually destroyed.

Application_Start is called when the application is requested from the server. This process in turn leads into the Event process(es). When all associated application modules have loaded, HTTPApplication.Init is called. The application executes its events, and when the user attempts to close it, Dispose is called. Dispose then passes processing on to the Application_End process, which closes the application.

This is a fairly standard application life cycle. Most applications follow similar life cycles: an application is created, loaded, has events, and is destroyed. The following section demonstrates how this compares to the Android application life cycle.

Android Application Life Cycle

The Android application life cycle is unique in that the system controls much of the life cycle of the application. All Android applications, or Activities, are run within their own process. All of the running processes are watched by Android and, depending on how the activity is running (this is, a foreground activity, background activity, and so forth), Android may choose to end the activity to reclaim needed resources.

NOTE

When deciding whether an activity should be shut down, Android takes into account several factors, such as user input, memory usage, and processing time.

Some of the specific methods called during the life cycle of an android activity are

- onCreate
- onStart
- Process-specific events (for example: launching activities or accessing a database)
- onStop
- onDestroy

Following the same logic as other application life cycles, an Android application is created, the processes are started, events are fired, processes are stopped, and the application is destroyed. Though there are a few differences, many application developers should be comfortable with the steps in the life cycle.

Ask the Expert

Q: Does Google ever update the Android SDK?

A: Yes. From the time I started writing this book, Google had already updated the Android SDK three times. Google will post the updates to the Android website as they are available.

Q: Do any of the API Demos represent applications that will be in the finished product?

A: Probably not. The API Demos were created to show off the capabilities of the product. Although there may be core "release" applications that contain some of the elements found in the API Demos, we probably will not see Lunar Lander in the finished version.

Chapter 5

Application: Hello World!

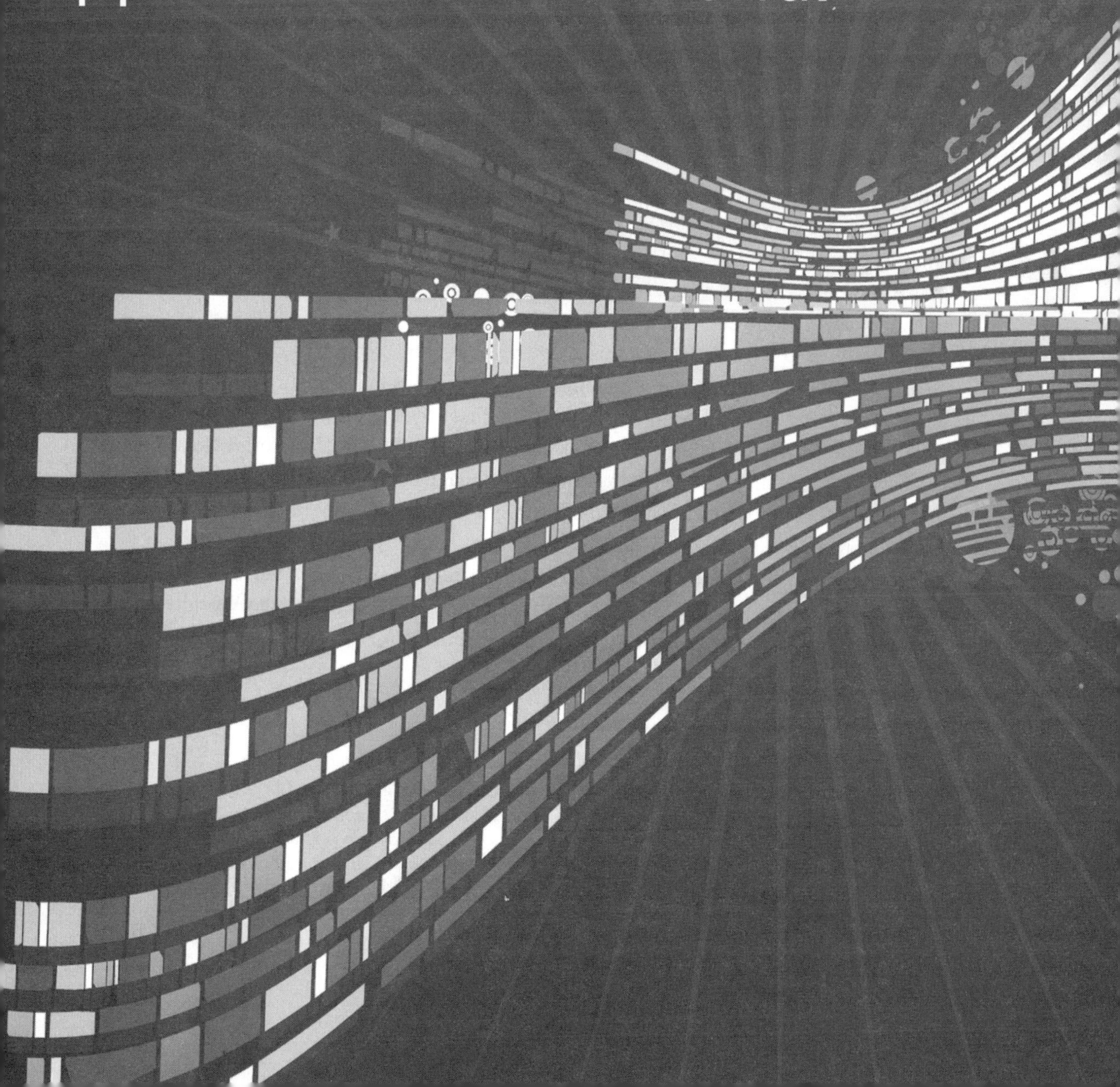

Key Skills & Concepts

- Creating new Android projects
- Working with Views
- Using a TextView
- Modifying the main.xml file
- Running applications on the Android Emulator

In this chapter, you will be creating your first Android Activity. This chapter examines the application-building process from start to finish. I will show you how to create an Android project in Eclipse, add code to the initial files, and run the finished application in the Android Emulator. The resulting application will be a fully functioning program running in an Android environment.

Actually, as you move through this chapter, you will be creating more than one Android Activity. Computer programming tradition dictates that your first application be the typical Hello World! application, so in the first section you will create a standard Hello World! application with just a blank background and the "Hello World!" text. Then, for the sake of enabling you to get to know the language better, the next section explains in detail the files automatically created by Android for your Hello World! application. You will create two iterations of this Activity, each using different techniques for displaying information to the screen. You will also create two different versions of a Hello World! application that will display an image that delivers the "Hello World!" message. This will give you a good introduction to the controls and inner workings of Android.

NOTE

You will often see "application" and "Activity" used interchangeably. The difference between the two is that an application can be composed of multiple Activities, but one application must have at least one Activity. Each "window" or screen of your application is a separate Activity. Therefore, if you create a fairly simple application with only one screen of data (like the Hello World! application in this chapter), that will be one Activity. In future chapters you will create applications with multiple Activities.

To make sure that you get a good overall look at programming in Android, in Chapter 6 you will create both of these applications in the Android SDK command-line environment for Microsoft Windows and Linux. In other words, this chapter covers the creation process in Eclipse, and Chapter 6 covers the creation process using the command-line tools. Therefore, before continuing, you should check that your Eclipse environment is correctly configured. Review the steps in Chapter 3 for setting the PATH statement for the Android SDK. You should also ensure that the JRE is correctly in your PATH statement.

TIP

If you have configuration-related issues while attempting to work with any of the command-line examples, try referring to the configuration steps in Chapters 2 and 3; and look at the Android SDK documentation.

Creating Your First Android Project in Eclipse

To start your first Android project, open Eclipse. When you open Eclipse for the first time, it opens to an empty development environment (see Figure 5-1), which is where you want to begin. Your first task is to set up and name the workspace for your application. Choose File | New | Android Project, which will launch the New Android Project wizard.

CAUTION

Do not select Java Project from the New menu. While Android applications are written in Java, and you are doing all of your development in Java projects, this option will create a standard Java application. Selecting Android Project enables you to create Android-specific applications.

If you do not see the option for Android Project, this indicates that the Android plugin for Eclipse was not fully or correctly installed. Review the procedure in Chapter 3 for installing the Android plugin for Eclipse to correct this.

The New Android Project wizard creates two things for you:

- A shell application that ties into the Android SDK, using the android.jar file, and ties the project into the Android Emulator. This allows you to code using all of the Android libraries and packages, and also lets you debug your applications in the proper environment.

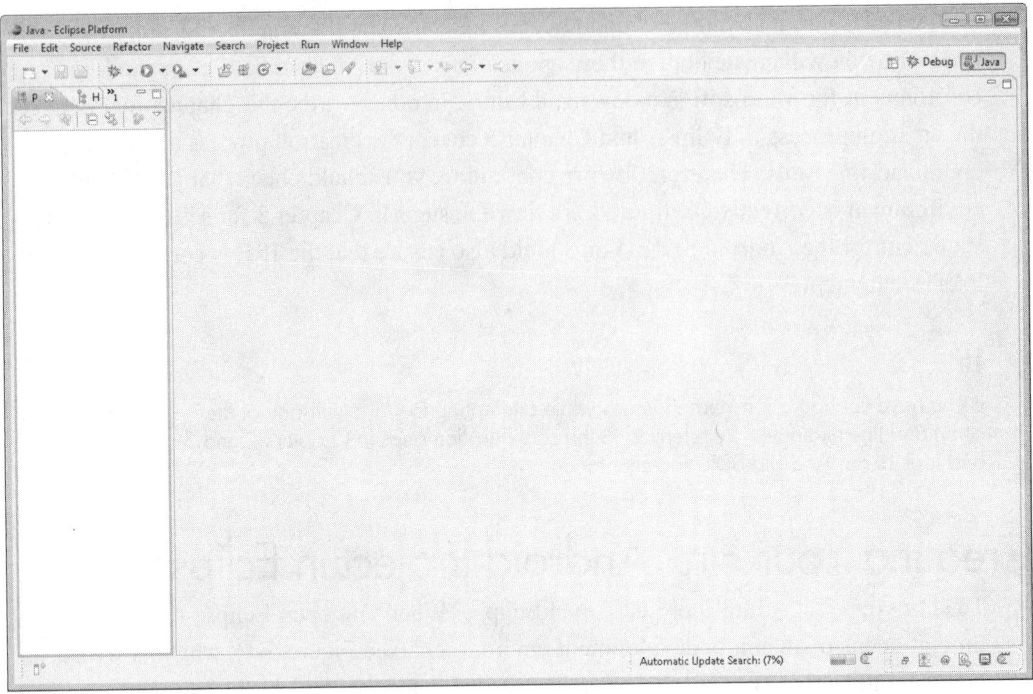

Figure 5-1 The empty Eclipse development environment

- Your first shell files for the new project. These shell files contain some of the vital application blocks upon which you will be building your programs. In much the same way as creating a Microsoft .NET application in Visual Studio generates some Windows-created program code in your files, using the Android Project wizard in Eclipse generates your initial program files and some Android-created code.

In addition, the New Android Project wizard contains a few options, shown next, that you must set to initiate your Android project.

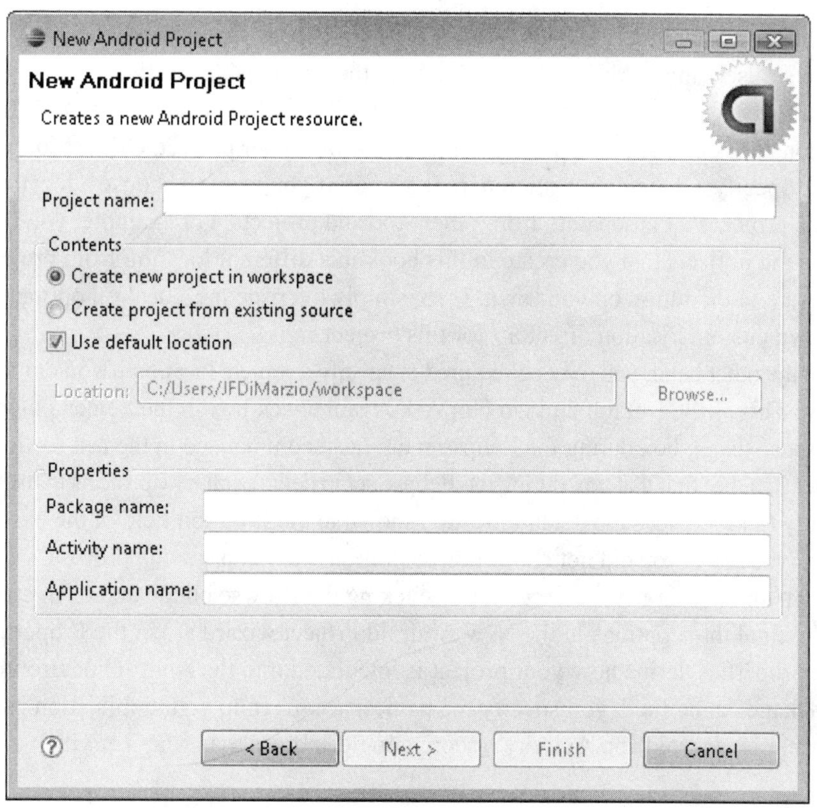

For the Project Name field, for purposes of this example, use the title **HelloWorldText**. This name sufficiently distinguishes this Hello World! project from the others that you will be creating in this chapter.

In the Contents area, keep the default selections: the Create New Project in Workspace radio button should be selected and the Use Default Location check box should be checked. This will allow Eclipse to create your project in your default workspace directory. The advantage of keeping the default options is that your projects are kept in a central location, which makes ordering, managing, and finding these projects quite easy. For example, if you are working in a Unix-based environment, this path points to your $HOME directory.

If you are working in a Microsoft Windows environment, the workspace path will be C:/Users/<username>/workspace, as shown in the previous illustration.

However, for any number of reasons, you may want to uncheck the Use Default Location check box and select a different location for your project. One reason you may want to specify a different location here is simply if you want to choose a location for this specific project that is separate from other Android projects. For example, you may want to keep the projects that you create in this book in a different location from projects that you create in the future on your own. If so, simply override the Location option to specify your own custom location directory for this project.

On the other hand, you may be required to specify a project location if you did not check the Use This as the Default and Do Not Ask Again check box in the Select a Default Workspace dialog box during the Eclipse setup (as recommended in the last section of Chapter 2). Checking that box during the Eclipse setup defaults all new projects to the workspace directory (and provides the default location shown in the Location field of the New Android Project wizard). If you did not check this box during the Eclipse setup process, you need to select a path for your new project now by clicking the Browse button and navigating to it.

The final three options in the New Android Project wizard are in the Properties area. These properties define how your project is integrated into the Android environment. In the Package Name field, you specify the namespace given to your application package. For example, android.app.Activity or com.google.android.map.MapActivity.

CAUTION

The package name adheres to the standard Java package-naming guidelines, which were established to lower the risk of two packages being released with the same name. The top level of the package name is the domain identifier of the company (com, org, and net are examples). This is followed by the domain name, such as google. Finally, a descriptive title for the contents of the package is provided. For purposes of this chapter, my package name for the Hello World! application will omit "com" to identify that it is a text application and not meant to be published. All future packages created in this book will be publishable and use the com identifier.

For the HelloWorldText application, use the package name **android_programmers_guide.HelloWorldText**. This name uniquely identifies the code that belongs to this application and differentiates this test application from others you will develop in this book.

CAUTION

If you are paying attention to the screen as you are typing, you will notice that an error message appears at the top of the wizard as you enter the package name, stating that you must fill out all the fields properly to continue. This error message is premature and can be a bit confusing because you have not even attempted to fill out the other fields in the Properties area. If you see such an error message, just ignore it and continue on and complete the next two fields in Properties area.

The next Properties field, Activity Name, is required because it is the reference to the main screen of your application. That is, think of the Activity as the "window" within which your application is displayed. Without an Activity, your application would not do very much. However, because Android applications can be composed of several Activities, the New Android Project wizard needs to know which Activity will be the default. Activity Name is a required field and has no default, so you must supply one to continue (as indicated in the preceding caution). For purposes of this example, use **HelloWorldText**. This keeps the application simple and is just about as descriptive as it needs to be for the moment.

The final Properties field, Application Name, specifies the name of your application. This is the name that will be used to manage your application when it is installed on the device. Again, for the sake of keeping things simple, go with **HelloWorldText** as the application name. The following illustration shows the completed New Android Project wizard.

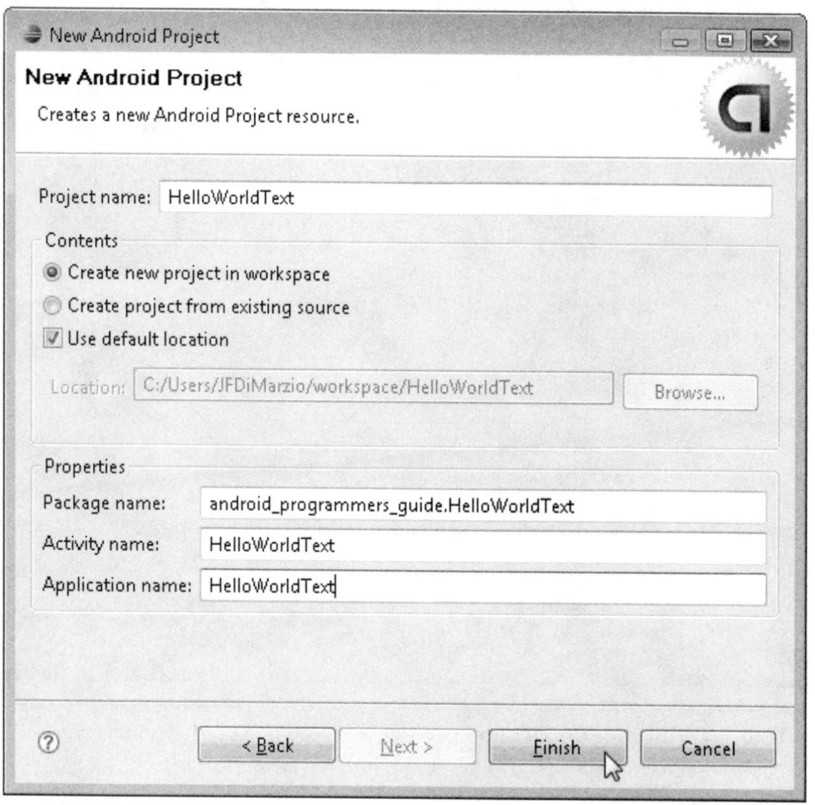

TIP

The Application Name and the Activity Name fields do not have to match. In fact, many programmers are used to the older conventions whereby the "starting" screen of an application is usually called Main or Startup. Use whatever conventions you are comfortable with. For purposes of demonstration, this chapter assumes that you are using the names suggested.

Click Finish to kick off the creation process. The wizard runs a background process that facilitates the auto-generation of some required files, and the setup of the directory structure needed to support an Android application. When the process is complete, you will have your first Android application project, like that shown in Figure 5-2.

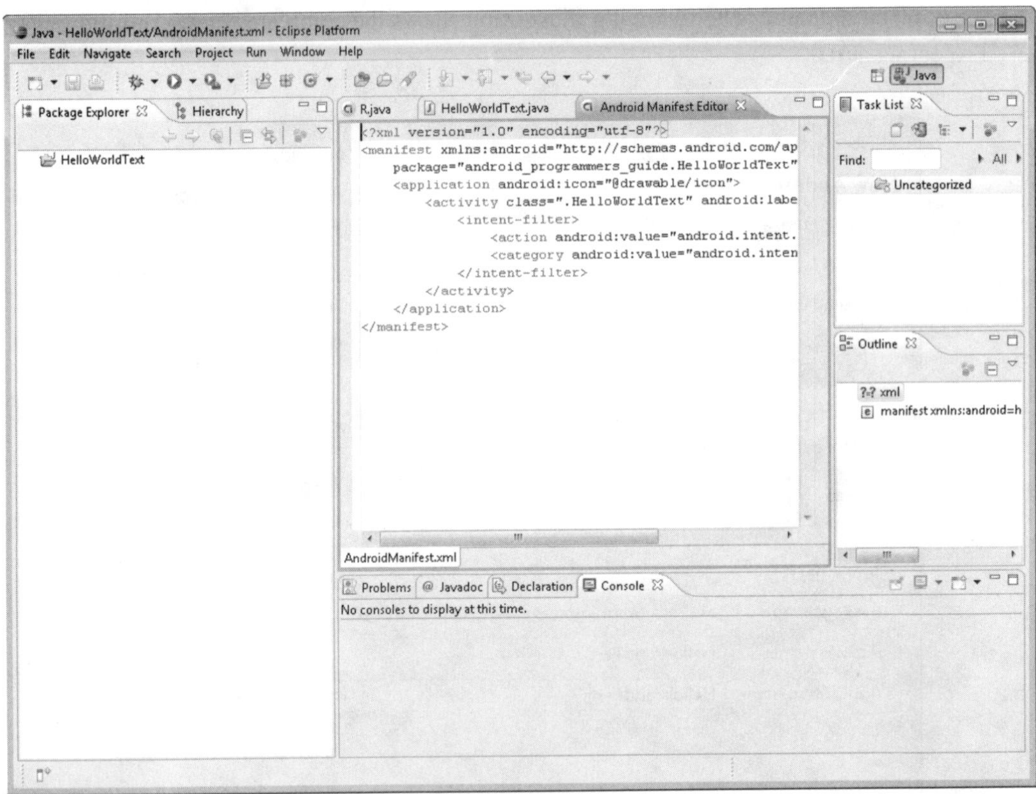

Figure 5-2 Your first Android application project

TIP

If the Finish button is not available to you, you may have made an error in one of the fields in the Properties area. To ensure that the Properties fields are correctly filled in, Eclipse will not allow you to finish the process if any of the information that you entered may cause problems. Go back and make sure that all of the Properties fields are correctly filled in.

The next section examines the contents of the auto-generated Android files and the purpose of some of the shell items for your application.

Examining the Android-Created Files

This section discusses the new files that Android has just created for you. A fairly robust structure has been created for you, and if you do not know what you are looking at, you may end up putting some code in places that you should not. There are files provided by Android that you need to modify, and there are ones that you should not modify; knowing the difference may save you from having to re-create your project.

With your new application project open, take a look at the Package Explorer, one of two tabs located in the pane to the left of the main development area. The following illustration shows what the Package Explorer should look like.

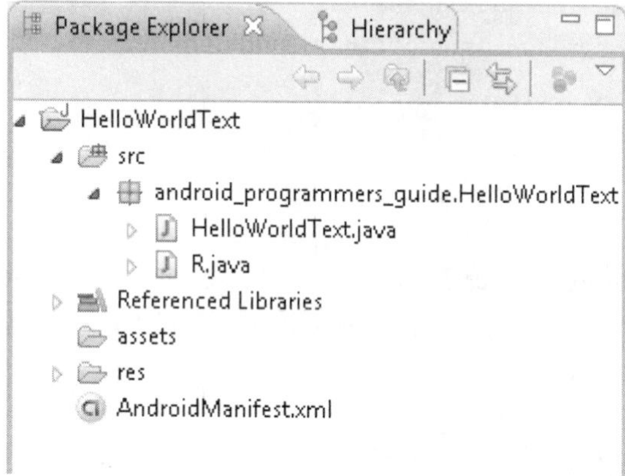

NOTE

If the Package Explorer is not open, you can activate it by choosing Window | Show View | Package Explorer.

You should see a root directory, in this case named HelloWorldText. The root directory is the home, or repository, for all of your project files. Both your user-created files and the Android auto-generated files will be placed in the directory, easily accessible from the Package Explorer. Currently there should be a few items in your root directory: an AndroidManifest.xml file, a package included in the Referenced Libraries, and three directories (res, assets, and src). These items are discussed in turn next.

AndroidManifest.xml

The AndroidManifest.xml file is where your global settings are made. If you are an ASP.NET developer, you can think of AndroidManifest.xml as Web.config and Global.asax rolled into one. (If you are not an ASP.NET developer, this means that AndroidManifest.xml is a place for storing settings.) AndroidManifest.xml will include such settings as application permissions, Activities, and intent filters.

The standard AndroidManifest.xml file should contain the following information:

```
<?xml version="1.0" encoding="utf-8"?>
<manifest xmlns:android=http://schemas.android.com/apk/res/android
    package="testPackage.HelloWorldText">
    <application android:icon="@drawable/icon">
        <activity class=".HelloWorldText" android:label="@string/app_name">
            <intent-filter>
                <action android:value="android.intent.action.MAIN" />
                <category android:value="android.intent.category.LAUNCHER"
/>
            </intent-filter>
        </activity>
    </application>
</manifest>
```

As you create future applications, you will be adding information to this file. Notice that the package name you supplied is listed here, as well as the action that your Activity will handle.

Referenced Libraries

A list of the Referenced Libraries is also included in the root of the project. Typically, for a beginner project, you should see only one library here. Expand the Referenced Libraries branch and examine its contents, the libraries that are currently referenced by your application project. Given that this is a new Android project, you will see one library in your project's references, android.jar, the Android SDK. (If you are familiar with the Java SDK, android.java is analogous to Java's rt.java file, containing many of the Java APIs found in rt.java.) The Android plugin ensures that this file is the only library referenced by your application. The application needs to reference the SDK to gain access to all the classes contained in the SDK libraries, such as your Views, Controls, and even the Google API.

CAUTION

Eclipse enables you to add other user-defined libraries and external classes to your project's references. However, unless you are sure that those external references will work with your Android application (and thus on the Android platform), you should think twice before you add them.

Directories

There are also three directories in the project root—res, assets, and src—each of which has a distinct purpose. These directories play an integral part in the operation of your application.

res Directory

The res directory is where your *in project* resources are held and compiled into your application. When you create a new Android project, the res directory contains three subdirectories: drawable, layout, and values. You will use the drawable and layout directories in many of your projects to hold and display images and layouts respectively, whereas the values directory holds string globals that can be used throughout your application.

NOTE

A reference to the res directory and its contents is contained by the R.java file, located in the src directory. This file is covered in much more detail later in the chapter.

The drawable directory contains actual image files that your application can use and reference. The layout directory holds an XML file, main.xml, that is referenced by your application when building its interface. In most of the applications in this book, you will be editing the main.xml file included in the layout directory. This will allow you to insert

Views into the application's visual layout and display them. An unaltered main.xml file contains the following code:

```
<?xml version="1.0" encoding="utf-8"?>
<LinearLayout xmlns:android=http://schemas.android.com/apk/res/android
android:orientation="vertical"
android:layout_width="fill_parent"
android:layout_height="fill_parent"
>
<TextView
    android:layout_width="fill_parent"
    android:layout_height="wrap_content"
    android:text="Hello World, HelloWorldText"
    />
</LinearLayout>
```

The last directory under res, values, holds an XML file named strings. The strings.xml file is used to hold global string values that can be referenced by your application.

assets Directory

The assets directory is used to hold raw asset files. The files contained in the assets directory can include audio files for streaming and animation assets. I will not use any audio assets in the applications for this book because the beta audio drivers for the Android Emulator are not yet optimized.

src Directory

The src directory contains all the source files for your project. When your project is first created, it will contain two files, R.java and <activity>.java (in this example, HelloWorldText.java), described next.

NOTE

<activity>.java is always named according to your Activity name.

R.java File The file R.java is an auto-generated file that is added to your application by the Android plugin. This file contains pointers into the drawable, layout, and values directories (or the items within the directories, as is the case with strings and icons). You should never have to modify this file directly. You will be referencing R.java in most of your applications. The code that was auto-generated for the HelloWorldText application follows:

```
/* AUTO-GENERATED FILE.  DO NOT MODIFY.
 *
 * This class was automatically generated by the
 * aapt tool from the resource data it found.  It
 * should not be modified by hand.
 */

package testPackage.HelloWorldText;

public final class R {
    public static final class attr {
    }
    public static final class drawable {
        public static final int icon=0x7f020000;
    }
    public static final class layout {
        public static final int main=0x7f030000;
    }
    public static final class string {
        public static final int app_name=0x7f040000;
    }
}
```

NOTE

The comment section of the R.java file provides an explanation of the origin of the file. It states that the file was created by the aapt tool. In Chapter 6, when you create a command-line–only version of the Hello World! application, you will use command-line tools to create all of the auto-generated files.

<activity>.java File The file in the src directory that you will spend the most time with is <activity>.java (HelloWorldText.java in this example), which is created by the Android plugin and named to match the Activity name that you specified in the New Android Project wizard. Unlike most of the files you have examined in this section, this file is completely editable; in fact, it will do very little for you if you do not modify it with your code.

After briefly looking at what is in your HelloWorldText.java file as it is created by the Android plugin, you will then edit the file to create your first Android Activity.

```
package android_programmers_guide.HelloWorldText;
import android.app.Activity;
import android.os.Bundle;
```

```
public class HelloWorldText extends Activity {
    /** Called when the activity is first created. */
    @Override
    public void onCreate(Bundle icicle) {
        super.onCreate(icicle);
        setContentView(R.layout.main);

    }
}
```

The three lines at the top of the file are the standard preprocessor directives—that is, as in most programming languages, statements that are directives to the compiler to run before the application process. In this case, you have the definition and inclusion of your package android_programmers_guide.HelloWorldText.

The next two lines import specific packages from the Android SDK via android.jar:

```
import android.app.Activity;
```

and

```
  import android.os.Bundle;
```

These lines tell the project to include all the code from the imported packages before all the code in your application. These two lines are critical for your base Android application and should not be removed.

TIP

If you do not see the android.os.Bundle import statement in your project, expand the tree within your development window. Eclipse rolls up all the import statements under the first one, so you must expand the tree to see the rest of them.

Focusing now on your class HelloWorldText, you can see that it extends the Activity class. Activity is imported from the previous lines. All applications derive the Activity class, and this derivation is required for running an application on Android. For something to run and be displayed on the screen, it must be derived from Activity.

The HelloWorldText class holds the code needed to create, display, and run your application. Right now there is only one method in your HelloWorldText class that is defined with code in it, onCreate( ).

The onCreate( ) method takes in `icicle` as a bundle. That is, all of the current state information is bundled as an `icicle` object and held in memory. You will not be directly handling `icicle` in this application, but you need to be aware of its presence and purpose.

The next line in the file is the one that really does some perceptible action:

```
setContentView(R.layout.main);
```

The method setContentView( ) sets the Activity's content to the specified resource. In this case, we are using the main.xml file from the layout directory via the pointer in the R.java file. The main.xml file, right now, contains nothing more than the size of the HelloWorldText screen and a TextView. The TextView is derived from View and is used to display text in an Android environment. Reviewing the contents of main.xml, you can see that it contains the following line:

```
android:text="Hello World, HelloWorldText"
```

Considering that the setContentView( ) method is being told to set main.xml as the current View, and main.xml contains a TextView that says "Hello World, HelloWorldText," it may be safe to assume that compiling and running HelloWorldText now will give you your Hello World! application. To test this, run your unaltered HelloWorldText application. Choose Run | Run to open the Run As dialog box, select Android Application, and click OK.

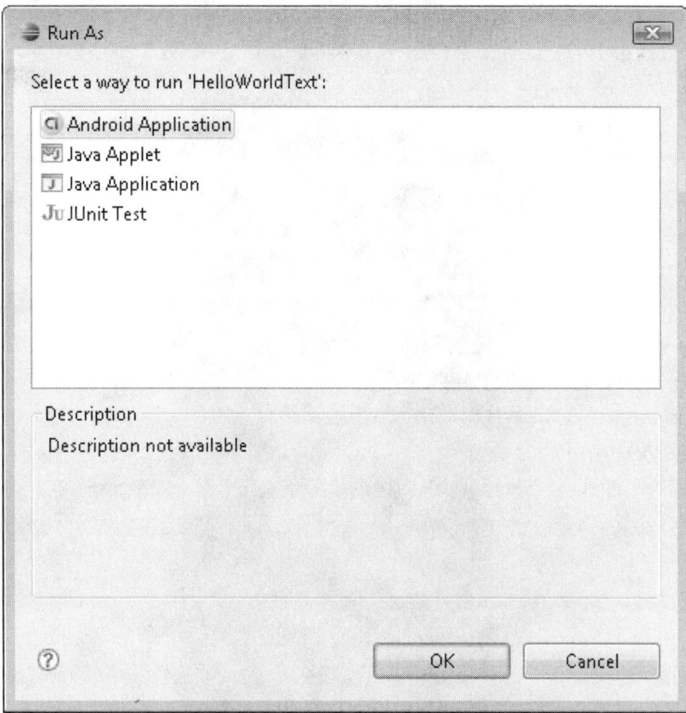

The new project you just established contains the code to create a Hello World! application on its own. However, that is not very engaging, nor does it teach you very much about programming an Android application. You need to dissect the project and see exactly how the project displayed the "Hello World!" message.

What happened when you created the new Android project is that the Android plugin modified main.xml. This is a perfect example of one way to modify the UI in Android. The following lines of code are added to main.xml by the Android SDK when the project is created:

```
<TextView
    android:layout_width="fill_parent"
    android:layout_height="wrap_content"
    android:text="Hello World, HelloWorldText"
/>
```

While I have discussed the existence of this TextView in the xml, I have not yet discussed why it works without any corresponding code. I mentioned earlier in this book that there are two ways to design a UI for Android: through the code, and through the main.xml file. The preceding code sample creates a TextView in xml and sets the text to "Hello World, HelloWorldText." Edit this line of the main.xml file to read as follows:

```
    android:text="This is the text of an Android TextView!"
```

Rerun the project, and your results should appear as they do in this illustration.

Take some time and experiment with the xml TextView. Then you can move on to another way of creating a Hello World! application.

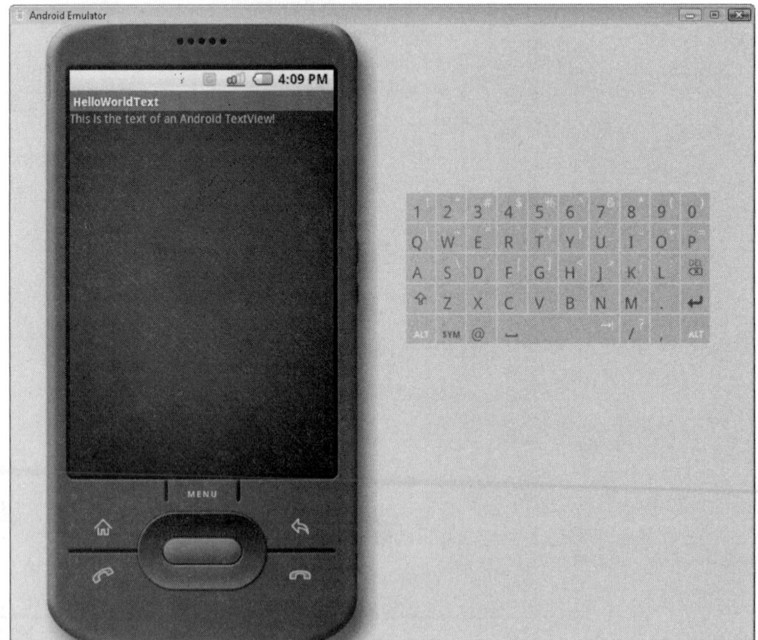

Hello World! Again

In this section, you will create another Hello World! application for Android. However, this time you will program the UI in code rather than by using the xml file—and you will actually do most of the work. The first step here is to remove the TextView code that is in main.xml. The following section of code represents the TextView. Removing it essentially makes your application an empty shell.

```
<TextView
    android:layout_width="fill_parent"
    android:layout_height="wrap_content"
    android:text="Hello World, HelloWorldText"
/>
```

After you have removed the TextView code, your main.xml file should look like this:

```
<?xml version="1.0" encoding="utf-8"?>
<LinearLayout xmlns:android=http://schemas.android.com/apk/res/android
    android:orientation="vertical"
    android:layout_width="fill_parent"
    android:layout_height="fill_parent"
    >

</LinearLayout>
```

Now that you have a clean main.xml file, and thus a clean application shell, you can begin to add the code that will display "Hello World!" on the screen. Start by opening the HelloWorldText.java file and removing the following line:

```
setContentView(R.layout.main);
```

NOTE

You still need to set a ContentView for your new application; however, you are going to implement it slightly differently from how it is implemented here, so it is best to just remove the entire statement for now.

This line uses setContentView( ) to draw the main.xml file to the screen. Since you will not be using main.xml to define your TextView, you will not be setting it to your view. Instead, you will be building the TextView in code.

Your next step is to import the package TextView from android.widget. This will give you access to the TextView and let you create your own instance of it. Place this code near the top of your current HelloWorldText.java file, where the existing import statements are

```
import android.widget.TextView;
```

Now, create an instance of TextView. By creating the TextView instance, you can use it to display text to the screen without directly modifying main.xml. Place the following code after the onCreate( ) statement is fired:

```
TextView HelloWorldTextView = new TextView(this);
```

NOTE

TextView takes a handle to the current context as an argument. Pass *this* to the TextView to associate it with the current context. If you follow the hierarchy through the SDK, HelloWorldText extends Activity, which extends ApplicationContext, which in turn extends Context. This is how you can pass *this* to your TextView.

The preceding line creates an instance of TextView named HelloWorldTextView and then instantiates HelloWorldTextView, by setting it to a *new TextView*. The new TextView is passed the context of *this* to be fully instantiated.

Now that the TextView is defined, you can add your text to it. The following line of code assigns the text "Hello World!" to the TextView:

```
HelloWorldTextView.setText("Hello World!");
```

This line lets you set the text of your TextView. setText( ) lets you assign a string to the TextView.

Your TextView has been created and now contains the message that you want to display. However, simply passing "Hello World!" to the TextView does not display anything to the screen. As discussed previously, you need to set the ContentView to display something to the screen. You have to use the following code to set TextView to the context and display it to the screen:

```
setContentView(HelloWorldTextView);
```

Examining this line, you can see that you pass to setContentView your TextView. The preceding three lines of code are what it takes to make your Hello World! application. You created a TextView, assigned your text to it, and set it to the screen. All things considered, this is not very complicated at all.

The full contents of your HelloWorldText.java file should look like the following:

```
package android_programmers_guide.HelloWorldText;

import android.app.Activity;
import android.os.Bundle;
import android.widget.TextView;

public class HelloWorldText extends Activity {
    /** Called when the activity is first created. */
    @Override
    public void onCreate(Bundle icicle) {
        super.onCreate(icicle);
        /**Hello World JFD */
        /**BEGIN          */
          /**Create TextView */
        TextView HelloWorldTextView = new TextView(this);
          /**Set text to Hello World */
        HelloWorldTextView.setText("Hello World!");
          /**Set ContentView to TextView */
        setContentView(HelloWorldTextView);
        /**END            */

    }
}
```

Now compile and run your new Hello World! application in the Android Emulator. Choose Run | Run or press CTRL-F11 to launch the application in the Android

Emulator. The following illustration depicts the results of your Hello World! application.

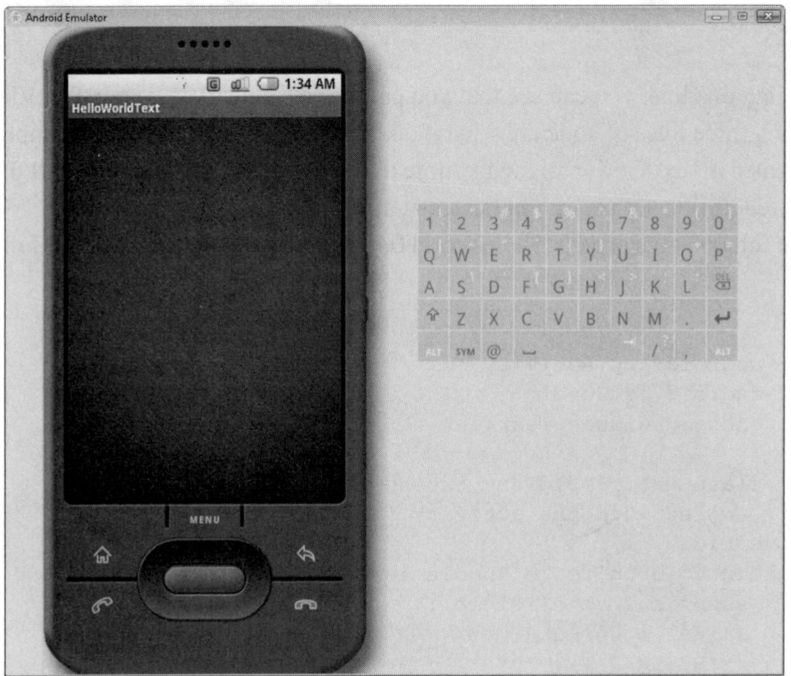

You have just created your first full Android Activity. This small project demonstrated a fairly common execution of a Hello World! application. You set a TextView to the Activity's ContentView and displayed the "Hello World!" message to a cell phone screen in the Android Emulator. The following section looks at a slightly different way of implementing Hello World!, using an image.

Hello World! Using an Image

In this section, you are going to use the Hello World! application to get more familiar with a relatively common practice in programming: displaying images. Modern computer displays would be exceedingly uninteresting without a graphical display. These graphical displays center on the ability to send images to the screen.

As late as five years ago, displaying images was a fairly difficult thing to do on a cell phone. Working with images is just one of those things that we, as modern PC users, take for granted. We look at windows of all types everyday without even considering that they are really images sent to a screen. This version of the Hello World! application will display an image to the screen that says "Hello World!"

For this application, use the New Android Project wizard to create a new project and name it **HelloWorldImage**, as shown in the following illustration.

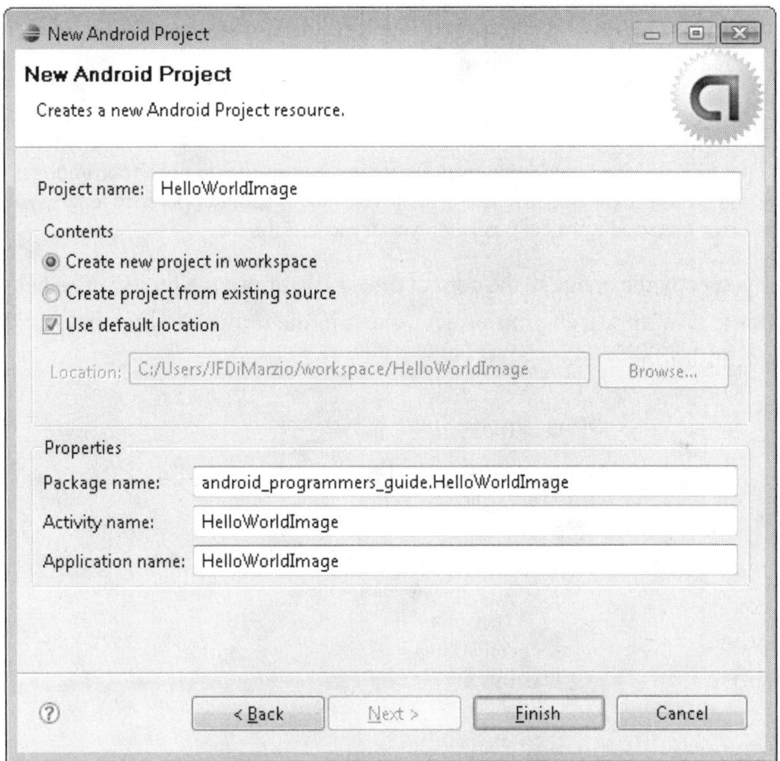

With the application project created, navigate to and remove the TextView code from main.xml so that you have a clean project. If you do not remove this code, you will end up with a text-based Hello World! program again.

Before you begin writing any code, you need an image to display. Create a small image in the graphics program of your choice. For ease of use, I chose Microsoft Paint, but any program should be able to give you the desired image. The image I am using is shown here:

Name your image **helloworld.png** and save it to the %workspace%/HelloWorldImage/res/drawable directory.

CAUTION

Be careful not to mix upper- and lowercase letters in your image names. Your images should be named using lowercase letters only. If you mix in some uppercase letters, you will get an error message from Eclipse when you try to use the file.

After you copy the image to the correct directory, refresh the project. The helloworld.png image should now appear in your project view, in the drawable directory, as shown in the following illustration.

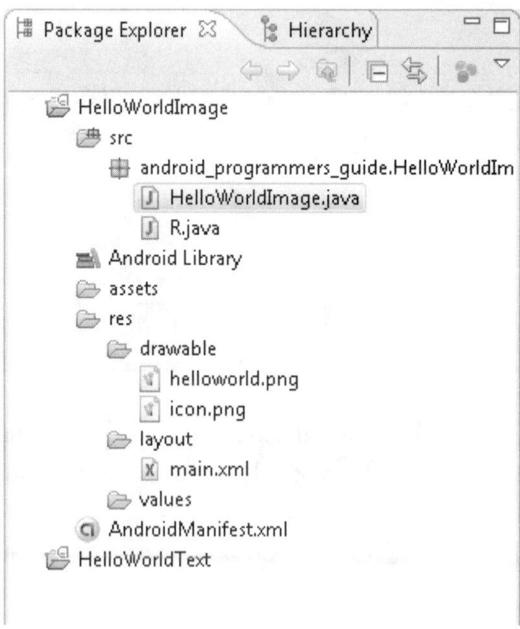

Open R.java and take a look at its code. Eclipse should have added a pointer to helloworld.png. Your R.java file should look similar to this:

```
/* AUTO-GENERATED FILE.  DO NOT MODIFY.
 *
 * This class was automatically generated by the
 * aapt tool from the resource data it found. It
 * should not be modified by hand.
 */

package android_programmers_guide.HelloWorldImage;

public final class R {
    public static final class attr {
    }
    public static final class drawable {
        public static final int helloworld=0x7f020000;
        public static final int icon=0x7f020001;
    }
    public static final class layout {
        public static final int main=0x7f030000;
    }
    public static final class string {
        public static final int app_name=0x7f040000;
    }
}
```

With a clean application shell as your starting point, and an available handle to the image you want to display, you can begin to add your code. You are going to look at this application from two perspectives: that of the XML-based UI and that of the code-based UI.

Hello World! Code-Based UI

Assuming you were able to follow along with and understand the HelloWorldText solution, this version of Hello World! will seem very familiar. To begin, you need to import the package for displaying images. Whereas text is displayed using a TextView, images are displayed using ImageView. Therefore, you must import the ImageView package. Like TextView, ImageView is contained in android.widgets:

```
import android.widgets.ImageView;
```

NOTE

Both TextView and ImageView are derived from View. This makes both of them very similar in structure and easy to implement.

With the package imported, you can create your ImageView and display it to the screen. Instantiating ImageView is the same as instantiating TextView; create an instance of ImageView and pass it the current context using *this*:

```
ImageView HelloWorldImageView = new ImageView(this);
```

The next line is where a difference between ImageView and TextView can be seen. This step involves setting your view to something that you want to display. In the TextView example, you used setText( ) to set the text of the TextView to "Hello World!" While both TextView and ImageView are derived from View, they are still different and therefore require some different methods. Obviously, you would not want to use setText( ) for your ImageView. You need to use setImageResource( ) to set the image in your ImageView. As shown next, pass into setImageResource( ) the handle to helloworld.png from R.java (the syntax for the handle is R.drawable.helloworld):

```
HelloWorldImageView.setImageResource(R.drawable.helloworld);
```

Finally, to send the image to the screen, you must set the ContentView. Just as you did with the TextView, you pass to the ContentView your ImageView. The job of the ContentView is then to set the object that it is passed to the screen.

```
setContentView(HelloWorldImageView);
```

Your final HelloWorldImage.java file should look like this:

```
package android_programmers_guide.HelloWorldImage;

import android.app.Activity;
import android.os.Bundle;
import android.widget.ImageView;

public class HelloWorldImage extends Activity {
    /** Called when the activity is first created. */
```

```
    @Override
    public void onCreate(Bundle icicle) {
        super.onCreate(icicle);
        /**Hello World Image JFD*/
        /**BEGIN                    */
        /**Create the ImageView */
        ImageView HelloWorldImageView = new ImageView(this);
        /**Set the ImageView to helloworld.png */
        HelloWorldImageView.setImageResource(R.drawable.helloworld);
        /**Set the ContentView to the ImageView */
        setContentView(HelloWorldImageView);
        /**END                               */

    }
}
```

Compile HelloWorldImage and run it in the Android Emulator. Your application should look similar to that in the following illustration.

In the next section, you will again display helloworld.png, but this time using XML rather than code.

Hello World! XML-Based UI

This section gives you a very good comparison by which to judge the processes of displaying images using the XML-based UI and the code-based UI. As you are going to see, the process of sending images to the screen using main.xml requires roughly the same amount of code as using the code-based UI. However, the syntax differs between the two processes.

Using the same project as you did for the last example, remove the TextView code from the HelloWorldImage.java file. The clean file should look like this:

```
package android_programmers_guide.HelloWorldImage;

import android.app.Activity;
import android.os.Bundle;

public class HelloWorldImage extends Activity {
    /** Called when the activity is first created. */
    @Override
    public void onCreate(Bundle icicle) {
        super.onCreate(icicle);

    }
}
```

Now that you have a clean slate to start with, move over to main.xml. You need to add in a definition for an ImageView. Start off by adding the empty ImageView tag to your main.xml file:

```
<ImageView
/>
```

You need to edit four attributes of the ImageView: android:id, android:layout_width, android:layout_height, and android:src. You are going to place these attributes in the tag, where they will govern how the tag is displayed to the screen.

The android:id attribute is set to the identifier for the ImageView. The android:id attribute can be used to refer to the ImageView in your code. Use the @+id/<*name*> syntax to assign to the ImageView an identity that can be retrieved later using R.layout.imageview:

```
android:id="@+id/imageview"
```

This line inserts an auto-generated ID, @+id, into the R.java file under the name imageview.

The next two attributes that you must define are android:layout_width and android:layout_height. These attributes govern how the image will fill the screen. There are two options you are going to select from when assigning values to these attributes. The fill_parent value fills the screen with the image while keeping it in perspective. The wrap_content value keeps the image its defined size, possibly losing some of the image definition in the process. For this example, use wrap_content:

```
  android:layout_width="wrap_content"
  android:layout_height="wrap_content"
```

The final attribute you need to assign is arguably the most important: android:src. This attribute points to the image that you want to display to the view. For this example, point the attribute to the drawable/helloworld image:

```
  android:src="@drawable/helloworld"
```

Your full ImageView tag should look like this:

```
<ImageView android:id="@+id/imageview"
  android:layout_width="wrap_content"
  android:layout_height="wrap_content"
  android:src="@drawable/helloworld"
  />
```

Finally, before the image will display to the view, you must pass main.xml to setContentView in HelloWorldImage.java:

```
setContentView(R.layout.main);
```

Compile and run HelloWorldImage. The results should look like the following illustration.

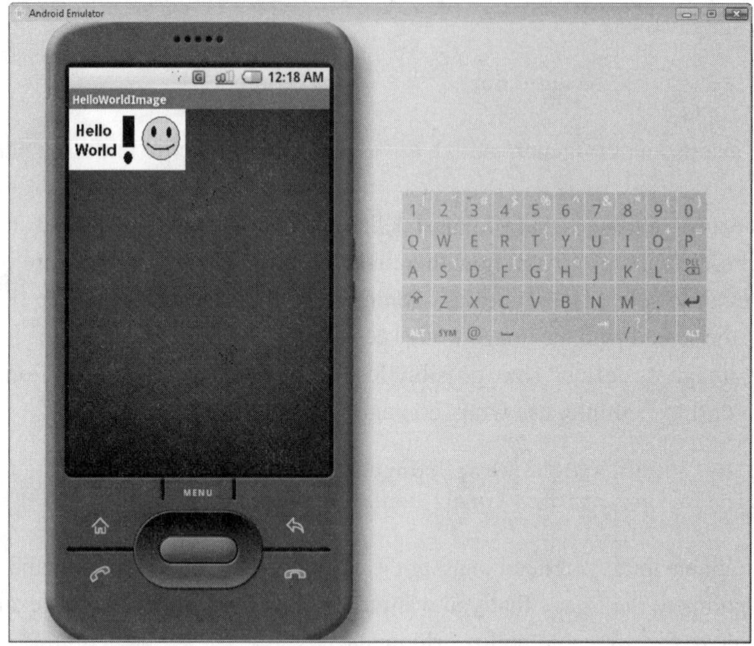

Before closing out this chapter, try one more thing. Go back to main.xml and change the layouts from wrap_content to fill_parent. When you are finished, your main.xml file should look like this:

```
<?xml version="1.0" encoding="utf-8"?>
<LinearLayout xmlns:android=http://schemas.android.com/apk/res/android
    android:orientation="vertical"
    android:layout_width="fill_parent"
    android:layout_height="fill_parent"
    >
<ImageView android:id="@+id/imageview"
  android:layout_width="fill_parent"
  android:layout_height="fill_parent"
  android:src="@drawable/helloworld"
  />
</LinearLayout>
```

Run the application again to see the difference between wrap_content and fill_parent. Your new application should look like the following illustration when it is run.

Try This Use TextView and ImageView

Use some of the skills and techniques that you learned in this chapter to create a new Hello World! application. Create an application that uses both the TextView and the ImageView to put an image on the screen with a text caption. This is slightly more difficult than using just one View on an Activity. Play with the Views and see what you can create.

The next chapter takes one more look at Hello World! applications, from the perspective of command-line programming.

Ask the Expert

Q: Does Android have a label or LabelView like most other APIs?

A: No. All text displays are facilitated through the TextView. You can, as some people have done, create a custom View that functions like a label and name it LabelView, but there is no packaged Android LabelView.

Q: Is there an advantage to using <application>.java rather than main.xml to create Views?

A: While there is no documented speed or processor savings in using one over the other, there is one key advantage: By using main.xml, you have a number of Views predefined for your Activity. Then, in your code, you can jump from View to View as needed without having to manually create them in code.

Chapter 6

Using the Command-Line Tools and the Android Emulator

Key Skills & Concepts

- Using the Android SDK command-line tools
- Creating a command environment
- Navigating the Android server with a shell
- Using the Android SDK in Linux

So far this book has covered some very broad subjects to get you up and running on the Android platform. At this point, you should be fairly comfortable using Eclipse to create and run a small Android application. You created a new project, edited the main.xml and <activity>.java files, and recompiled the R.java file. These are the basic skills that you need to create Android applications.

In this chapter, you are going to expand and round out those skills by experimenting with command-line application development. Android development does not have to be limited to the confines of the Eclipse IDE. The Android SDK offers a host of command-line tools that can help you develop full applications without the need of a graphical IDE. You will use these command-line tools to create, compile, and run a Hello World! application, first in Windows and then in Linux.

Creating a Shell Activity Using the Windows CLI

The Android SDK comes with multiple tools to help you create and compile Android applications. These tools are in place to help users who do not wish to, or do not have a system capable of supporting, work within a GUI IDE. However, if you are doing all of your Android development work within Eclipse, you still should be aware of the Android SDK command-line tools and their functionality.

When you run Android-related functions, such as creating an Android project or running an application in the Android Emulator, you are actually calling connections to the Android command-line tools. These Android command-line tools, whether run from a command-line interface or from a GUI IDE, are the real core to the functionality of the Android SDK.

In the following section, I demonstrate the functionality of one Android tool. The ActivityCreator.bat is a powerful tool that is used to establish shells of Activities that are ready for you to program.

Running the ActivityCreator.bat

The ActivityCreator.bat should be located in the ../tools/ directory of the Android SDK. Most of the forward-facing command-line tools are located in the root of the tools directory. "Forward-facing" tools are tools that in turn call other tools located deeper in the ../tools/ directory. ActivityCreator.bat is an example of a tool from the root of the tools directory that actually calls another tool when it is run. Using vi, Notepad, or any text editor, open ActivityCreator.bat; it should contain the following lines of code:

NOTE

ActivityCreator.bat is specific to the Microsoft Windows version of the Android SDK. In a later section of this chapter you will also learn about the ActivityCreator.py. This is the Linux version of the ActivityCreator.

```
@echo off
rem Copyright (C) 2007 Google Inc.
rem
rem Licensed under the Apache License, Version 2.0 (the "License");
rem you may not use this file except in compliance with the License.
rem You may obtain a copy of the License at
rem
rem      http://www.apache.org/licenses/LICENSE-2.0
rem
rem Unless required by applicable law or agreed to in writing, software
rem distributed under the License is distributed on an "AS IS" BASIS,
rem WITHOUT WARRANTIES OR CONDITIONS OF ANY KIND, either express or implied.
rem See the License for the specific language governing permissions and
rem limitations under the License.

rem don't modify the caller's environment
setlocal

"%~dp0\lib\activityCreator\activityCreator.exe" %*
```

Navigating through all of the rem statements (batch file comment statements), you will see that there is one line of practical code at the bottom of the file. ActivityCreator.bat is used to call ActivityCreator.**exe** in the ../tools/lib/activityCreator/ directory. The ActivityCreator.bat is an example of a tool that is really just a front end of other tools in the SDK.

So, what does ActivityCreator.bat (or ActivityCreator.exe) do? ActivityCreator is used to establish your development environment to the point where it will create the initial files needed to begin developing your application. This directory structure is the same structure discussed in Chapter 5. ActivityCreator.bat creates R.java, AndroidManifest.xml, and all the supporting files needed to begin your application.

Let's now go to a command-line environment and explore the ActivityCreator.

From your Start menu, click Run, type **CMD** or **COMMAND** in the Run dialog box, and click OK.

Executing this command launches the command window shown next. This window is the equivalent of the older DOS operating environments.

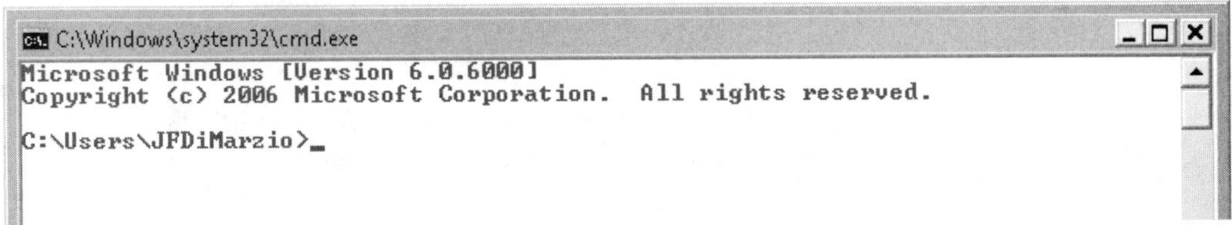

With the command window open, enter **ActivityCreator** at the > prompt.

TIP

The Microsoft command-prompt interface is not case sensitive, by default. If you are using a different environment that is case sensitive, such as Linux/Unix, the screenshots in this chapter may not show the proper case for your environment.

Running the command ActivityCreator, which actually runs ActivityCreator.bat, produces the following output:

```
Activity Creator Script
Usage:
  activityCreator [--out outdir] [--ide intellij] yourpackagename.ActivityName


Creates the structure of a minimal Android application.
The following will be created:
  - AndroidManifest.xml: The application manifest file.
  - build.xml: An Ant script to build/package the application.
  - res : The resource directory.
  - src : The source directory.
```

```
  - src/your/package/name/ActivityName.java the Activity java class. packageName
is a fully qualified java Package in the format <package1>.<package2>... (with at
least two components).
  - bin : The output folder for the build script.

Options:
 --out <folder>: specifies where to create the files/folders.
 --ide intellij: creates project files for IntelliJ
```

This output simply indicates that you need to provide more information to run ActivityCreator. More specifically, you need to pass the command a directory location within which to build your shell application.

NOTE

The output from the ActivityCreator command gives you a lot more information than just the fact that you did not provide enough information. It gives the complete list of the files created when using the tool. This list should look similar to that covered in Chapter 5, although build.xml is not directly exposed to Eclipse users.

Go back to the command window and run ActivityCreator with the following option (ActivityCreator also accepts path parameters with Unix-style forward slashes, if you are used to programming in a Unix/Linux environment):

```
--out c:\AndroidHelloWorld\android_programmers_guide.HelloWorldCommandLine
```

The --out option tells ActivityCreator that you want it to output something. This command option takes two parameters, <folder> and <package.activity>. The first part of the preceding line tells ActivityCreator to build the shell application in the nonexistent folder c:\AndroidHelloWorld.

TIP

If you specify a folder or directory that does not exist, ActivityCreator will create it for you during its process.

The second parameter of the --out option is the package name and the activity name. Following the convention established in the previous chapter, this example uses android_programmers_guide as the package name and HelloWorldCommandLine as the activity name for this project.

NOTE

The parameters needed to successfully run ActivityCreator and set up your initial environment are the same as those required by the New Android Project wizard.

Run ActivityCreator with this new command-line option and its parameter. You should see the following output from the tool:

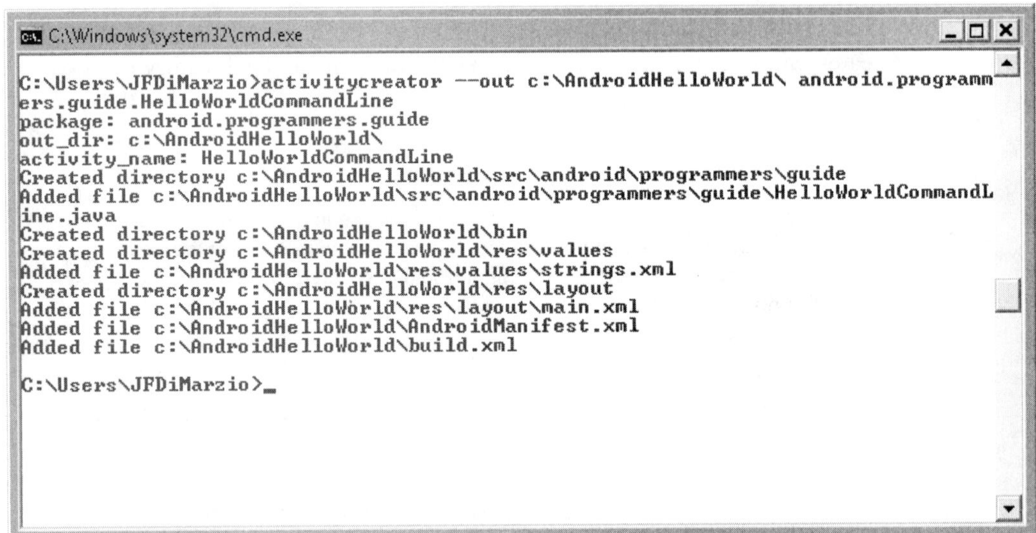

C:\Windows\system32\cmd.exe
C:\Users\JFDiMarzio>activitycreator --out c:\AndroidHelloWorld\ android.programm
ers.guide.HelloWorldCommandLine
package: android.programmers.guide
out_dir: c:\AndroidHelloWorld\
activity_name: HelloWorldCommandLine
Created directory c:\AndroidHelloWorld\src\android\programmers\guide
Added file c:\AndroidHelloWorld\src\android\programmers\guide\HelloWorldCommandL
ine.java
Created directory c:\AndroidHelloWorld\bin
Created directory c:\AndroidHelloWorld\res\values
Added file c:\AndroidHelloWorld\res\values\strings.xml
Created directory c:\AndroidHelloWorld\res\layout
Added file c:\AndroidHelloWorld\res\layout\main.xml
Added file c:\AndroidHelloWorld\AndroidManifest.xml
Added file c:\AndroidHelloWorld\build.xml
C:\Users\JFDiMarzio>_

The following section covers the files created by ActivityCreator, because they do vary slightly from those created by Eclipse.

The Project Structure

ActivityCreator created a number of directories and files for you to use and begin your development. Navigate to the c:\AndroidHelloWorld\ directory to explore its structure. ActivityCreator created the structure shown in the following illustration.

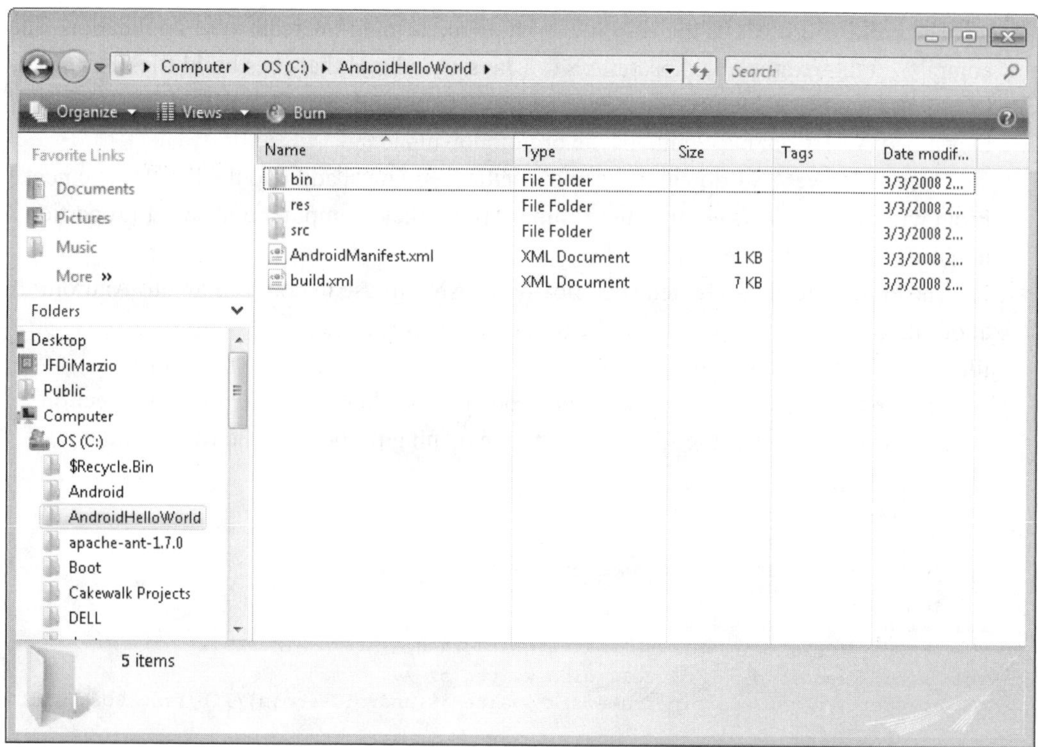

Because you are working outside of the Eclipse environment, you have a slightly different environment. When you are working within an IDE such as Eclipse, certain functions are performed behind the scenes for you. Given that you are working without any IDE help, ActivityCreator creates a file that outlines what the complier needs to do to create your project. ActivityCreator, when run manually, creates the build.xml file for your project. This file is not created when you use Eclipse to begin an Android project. It contains an instruction set that explains how to turn your .java files into a functional Android project.

The build.xml file tells the compiler what it needs to do to create your application. The compiler in this example is Apache ANT, a Java-based tool that uses build files as scripts to compile projects. You need to download ANT to compile your command-line project. Download ANT from http://ant.apache.org/bindownload.cgi.

Once you have ANT downloaded and installed, you must add it to the PATH statement. From within a Windows environment, simply right-click Computer and select Properties to change the PATH statement.

The build.xml file is created specifically for ANT to use in compiling your Android application. It should be located in the root of your project, as shown in the previous illustration. Open build.xml with your text editor and take a look at what is inside.

The first section of build.xml contains code that is editable by the user. This section is set off from the rest of the file because the remaining portions of the file should not be modified.

```
<?xml version="1.0" ?>
<project name="HelloWorldCommandLine" default="package">
    <property name="sdk-folder" value="c:\Android\android-sdk_m5
rc14_windows\android-sdk_m5-rc14_windows" />
    <property name="android-tools" value="c:\Android\android-sdk_m5
rc14_windows\android-sdk_m5-rc14_windows\tools" />
    <property name="android-framework" value="${android-tools}/lib/framework.aidl"
/>

    <!-- The intermediates directory -->
    <!-- Eclipse uses "bin" for its own output, so we do the same. -->
    <property name="outdir" value="bin" />
```

The first section of build.xml contains values for the following properties:

- Project name
- Android SDK location
- Android tools location
- Android framework location
- Output location

If you need to change any of these parameters for your project, you can do so within this file. However, immediately following these parameters in build.xml, you should see a warning informing you that you should not edit any remaining values:

```
<!-- No user servicable parts below. -->
```

Following this warning in build.xml is a list of parameters and values that are critical to the proper creation of your project. This list includes compiler options, input directories, and tool locations. Take a look at the following output of the core processing information of build.xml:

NOTE

While Android advises against changing the following parameters, if you are very familiar with how ANT works, you can modify these options to suit a particular need you may have.

```
<!-- Input directories -->
<property name="resource-dir" value="res" />
<property name="asset-dir" value="assets" />
<property name="srcdir" value="src" />

<!-- Output directories -->
<property name="outdir-classes" value="${outdir}/classes" />

<!-- Create R.java in the source directory -->
<property name="outdir-r" value="src" />

<!-- Intermediate files -->
<property name="dex-file" value="classes.dex" />
<property name="intermediate-dex" value="${outdir}/${dex-file}" />

<!-- The final package file to generate -->
<property name="out-package" value="${outdir}/${ant.project.name}.apk"/>

<!-- Tools -->
<property name="aapt"        value="${android-tools}/aapt" />
<property name="aidl"        value="${android-tools}/aidl" />
<property name="dx"          value="${android-tools}/dx" />
<property name="adb"         value="${android-tools}/adb" />
<property name="android-jar" value="${sdk-folder}/android.jar" />
```

```
<property name="zip"        value="zip" />

<!-- Rules -->

<!-- Create the output directories if they don't exist yet. -->
<target name="dirs">
    <mkdir dir="${outdir}" />
    <mkdir dir="${outdir-classes}" />
</target>

<!-- Generate the R.java file for this project's resources. -->
<target name="resource-src" depends="dirs">
    <echo>Generating R.java...</echo>
    <exec executable="${aapt}" failonerror="true">
        <arg value="compile" />
        <arg value="-m" />
        <arg value="-J" />
        <arg value="${outdir-r}" />
        <arg value="-M" />
        <arg value="AndroidManifest.xml" />
        <arg value="-S" />
        <arg value="${resource-dir}" />
        <arg value="-I" />
        <arg value="${android-jar}" />
    </exec>
</target>

<!-- Generate java classes from .aidl files. -->
<target name="aidl" depends="dirs">
    <apply executable="${aidl}" failonerror="true">
       <arg value="-p${android-framework}" />
        <arg value="-I${srcdir}" />
        <fileset dir="${srcdir}">
            <include name="**/*.aidl"/>
        </fileset>
    </apply>
</target>

<!-- Compile this project's .java files into .class files. -->
<target name="compile" depends="dirs, resource-src, aidl">
    <javac encoding="ascii" target="1.5" debug="true" extdirs=""
            srcdir="."
            destdir="${outdir-classes}"
            bootclasspath="${android-jar}" />
</target>
<!-- Convert this project's .class files into .dex files. -->
```

```
<target name="dex" depends="compile">
    <exec executable="${dx}" failonerror="true">
        <arg value="-JXmx384M" />
        <arg value="--dex" />
        <arg value="--output=${basedir}/${intermediate-dex}" />
        <arg value="--locals=full" />
        <arg value="--positions=lines" />
        <arg path="${basedir}/${outdir-classes}" />
    </exec>
</target>

<!-- Put the project's resources into the output package file. -->
<target name="package-res-and-assets">
    <echo>Packaging resources and assets...</echo>
    <exec executable="${aapt}" failonerror="true">
        <arg value="package" />
        <arg value="-f" />
        <arg value="-c" />
        <arg value="-M" />
        <arg value="AndroidManifest.xml" />
        <arg value="-S" />
        <arg value="${resource-dir}" />
        <arg value="-A" />
        <arg value="${asset-dir}" />
        <arg value="-I" />
        <arg value="${android-jar}" />
        <arg value="${out-package}" />
    </exec>
</target>

<!-- Same as package-res-and-assets, but without "-A ${asset-dir}" -->
<target name="package-res-no-assets">
    <echo>Packaging resources...</echo>
    <exec executable="${aapt}" failonerror="true">
        <arg value="package" />
        <arg value="-f" />
        <arg value="-c" />
        <arg value="-M" />
        <arg value="AndroidManifest.xml" />
        <arg value="-S" />
        <arg value="${resource-dir}" />
        <!-- No assets directory -->
        <arg value="-I" />
        <arg value="${android-jar}" />
        <arg value="${out-package}" />
    </exec>
</target>
```

```
<!-- Invoke the proper target depending on whether or not
     an assets directory is present. -->
<!-- TODO: find a nicer way to include the "-A ${asset-dir}" argument
     only when the assets dir exists. -->
<target name="package-res">
    <available file="${asset-dir}" type="dir"
            property="res-target" value="and-assets" />
    <property name="res-target" value="no-assets" />
    <antcall target="package-res-${res-target}" />
</target>

<!-- Put the project's .class files into the output package file. -->
<target name="package-java" depends="compile, package-res">
    <echo>Packaging java...</echo>
    <jar destfile="${out-package}"
            basedir="${outdir-classes}"
            update="true" />
</target>

<!-- Put the project's .dex files into the output package file.
     Use the zip command, available on most unix/Linux/MacOS systems,
     to create the new package (Ant 1.7 has an internal zip command,
     however Ant 1.6.5 lacks it and is still widely installed.)
-->
<target name="package-dex" depends="dex, package-res">
    <echo>Packaging dex...</echo>
    <exec executable="${zip}" failonerror="true">
        <arg value="-qj" />
        <arg value="${out-package}" />
        <arg value="${intermediate-dex}" />
    </exec>
</target>

<!-- Create the package file for this project from the sources. -->
<target name="package" depends="package-dex" />

<!-- Create the package and install package on the default emulator -->
<target name="install" depends="package">
    <echo>Sending package to default emulator...</echo>
    <exec executable="${adb}" failonerror="true">
        <arg value="install" />
        <arg value="${out-package}" />
    </exec>
</target>

</project>
```

Now that you have a good understanding of how build.xml is used in the manual, command-line creation of Android projects, you can begin to edit your project files and create an Android Activity. The first file you need to look at is main.xml. Using Windows Explorer, navigate to main.xml at AndroidHelloWorld\res\layout.

Creating the Hello World! Activity in the Windows CLI

In this section you will use the Windows command-line interface to edit the project files. The project files were created in the previous sections by the ActivityCreator.bat. You will edit these files and add code to them, without using Eclipse.

Editing the Project Files

Open main.xml in either an XML editor or (if you do not have a specific XML editor) Notepad. This will let you edit the file and remove the <TextView/> definition that is within it. Save main.xml as shown in the next illustration.

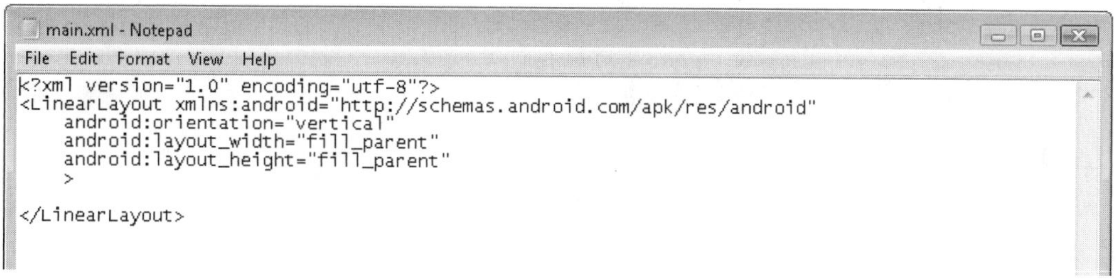

main.xml - Notepad

File Edit Format View Help

```
<?xml version="1.0" encoding="utf-8"?>
<LinearLayout xmlns:android="http://schemas.android.com/apk/res/android"
    android:orientation="vertical"
    android:layout_width="fill_parent"
    android:layout_height="fill_parent"
    >

</LinearLayout>
```

The result of saving main.xml is an empty shell. This gives you a platform on which to edit your <activity>.java file. The <activity>.java file is in a folder that is several directories deep, AndroidHelloWorld\src\android\programmers\guide.

To create your Hello World! application, add the following lines to create, set, and use a TextView:

```
/**Hello World JFD */
/**BEGIN          */
  /**Create TextView */
TextView HelloWorldTextView = new TextView(this);
  /**Set text to Hello World */
HelloWorldTextView.setText("Hello World!");
  /**Set ContentView to TextView */
setContentView(HelloWorldTextView);
/**END            */
```

Do not forget to add the TextView package to the beginning of the file:

```
import android.widget.TextView;
```

The finished HelloWorldCommandLine.java file should look like that in the following illustration.

```
HelloWorldCommandLine.java - Notepad
File Edit Format View Help
package android.programmers.guide;

import android.app.Activity;
import android.os.Bundle;
import android.widget.TextView;

public class HelloWorldCommandLine extends Activity
{
    /** Called when the activity is first created. */
    @Override
    public void onCreate(Bundle icicle)
    {
        super.onCreate(icicle);
        setContentView(R.layout.main);
        /**Hello World JFD */
        /**BEGIN          */
          /**Create TextView /
        TextView HelloWorldTextView = new TextView(this);
          /**Set text to Hello World /
        HelloWorldTextView.setText("Hello World!");
          /**Set ContentView to TextView /
        setContentView(HelloWorldTextView);
        /**END            */

    }
}
```

Your project files should now be set. You can now compile your program and run it in the Android Emulator.

Adding the JAVA_HOME Variable

Before you try compiling your project, you must add another environment variable to your PC—JAVA_HOME—that points to your JDK. Even if this path is in your PATH statement, you still need to create a new variable named JAVA_HOME.

NOTE

The JAVA_HOME variable is needed only if you are working in the command-line environment. If you are exclusively using Eclipse, you do not need to add it.

1. Right-click My Computer and select Properties.
2. Select the Advanced tab on the System Properties window and click the Environment Variables button. This will open an Environment Variables window.
3. Click the New button to add a new variable named JAVA_HOME. The value for the variable should be the full path to your Java SDK, as seen in the following illustration.

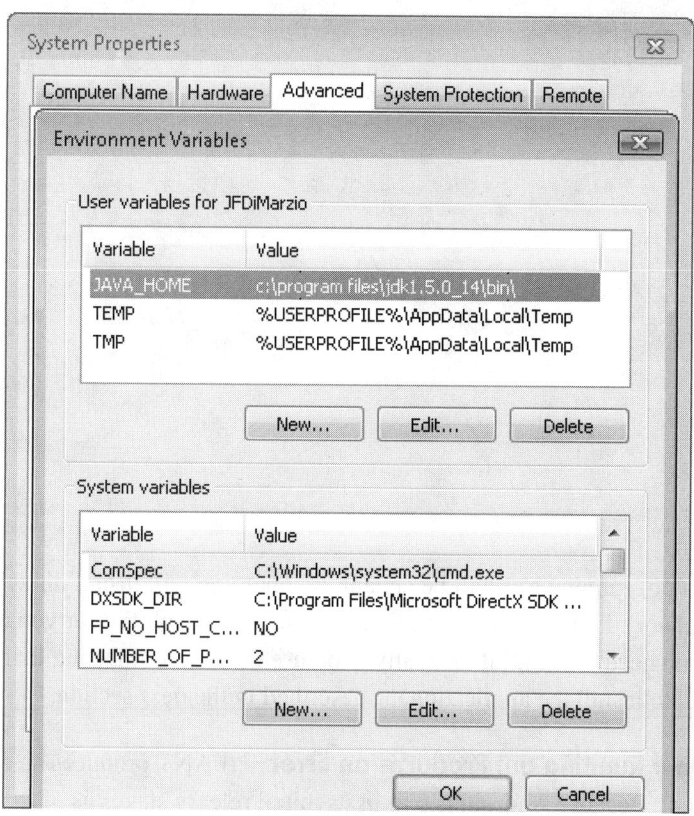

Compiling and Installing the Application

It is time for the real test. You can now compile your command-line project. To compile your project, use ANT. Once the project is compiled, you will install it on your Emulator.

Compiling Your Project with ANT

After you have your JAVA_HOME environment variable set and have ANT in your PATH statement, you should be able to navigate the directory containing your build.xml file and simply run the command **ant**. Open a Windows command prompt to your project directory and run **ant**, as follows:

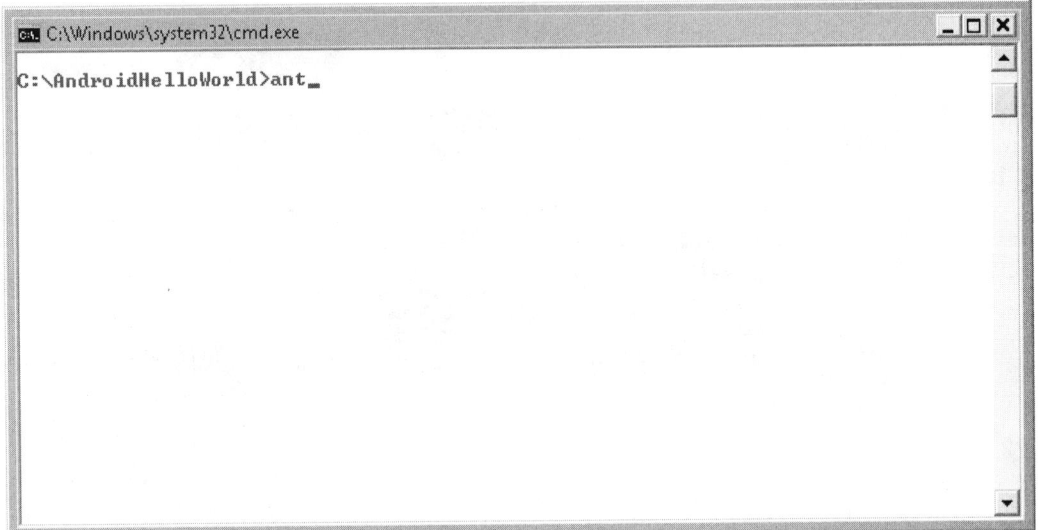

The result of running **ant** will be an .apk file that you will install directly onto your phone (Emulator). However, whereas Eclipse installs the .apk file for you directly on the Emulator, you need to install it manually. You use the Android Debug Bridge (adb) Android tool to install the application, as described in the next section.

What to Do if Running ant Produces an Error If ANT produces an error when you run it, fear not. Because Android is still in its initial release stages as of the writing of this book, several items may need to be tweaked. Small changes here and there can always be expected when you are working in a new technology. When I first tried to run **ant** and compile my project, I received an error like that shown in the following illustration.

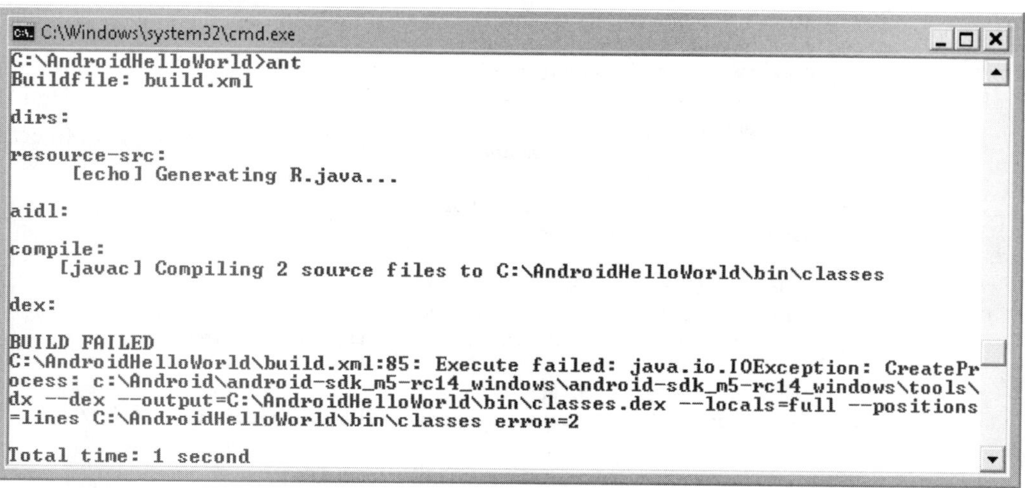

C:\Windows\system32\cmd.exe
C:\AndroidHelloWorld>ant
Buildfile: build.xml
dirs:
resource-src:
[echo] Generating R.java...
aidl:
compile:
[javac] Compiling 2 source files to C:\AndroidHelloWorld\bin\classes
dex:
BUILD FAILED
C:\AndroidHelloWorld\build.xml:85: Execute failed: java.io.IOException: CreatePr
ocess: c:\Android\android-sdk_m5-rc14_windows\android-sdk_m5-rc14_windows\tools\
dx --dex --output=C:\AndroidHelloWorld\bin\classes.dex --locals=full --positions
=lines C:\AndroidHelloWorld\bin\classes error=2
Total time: 1 second

Some research on the issue at the Google Android developer's forum turned up an interesting solution: a rewrite of build.xml that tweaked some of the commands offered to ANT. What follows is the modified build.xml file, in which the key changes have been bolded. Compare this file with the original and you will see that it differs quite noticeably.

```
<?xml version="1.0" ?>
<project name="HelloWorldCommandLine" default="package" basedir=".">
    <property name="sdk-folder" value="c:\Android\android-sdk_m5
rc14_windows\android-sdk_m5-rc14_windows" />
    <property name="android-tools" value="c:\Android\android-sdk_m5
rc14_windows\android-sdk_m5-rc14_windows\tools" />
    <property name="android-framework" value="${android-tools}/lib/framework.aidl"
/>

    <!-- The intermediates directory -->
    <!-- Eclipse uses "bin" for its own output, so we do the same. -->
    <!--  Use full path for output dir - FIX - BLOCK START  -->
    <property name="outdir" value="${basedir}/bin" />
    <!-- Use full path for output dir - FIX - BLOCK END -->

    <!-- No user servicable parts below. -->

    <!-- Input directories -->
```

```
    <property name="resource-dir" value="res" />
    <property name="asset-dir" value="assets" />
    <property name="srcdir" value="src" />

    <!-- Output directories -->
    <property name="outdir-classes" value="${outdir}/classes" />

    <!-- Create R.java in the source directory -->
    <property name="outdir-r" value="src" />

    <!-- Intermediate files -->
    <property name="dex-file" value="classes.dex" />
    <property name="intermediate-dex" value="${outdir}/${dex-file}" />

    <!-- The final package file to generate -->
    <property name="out-package" value="${outdir}/${ant.project.name}.apk"/>

    <!-- Tools -->
    <property name="aapt" value="${android-tools}/aapt" />
    <property name="aidl" value="${android-tools}/aidl" />

     <condition property="dx" value="${android-tools}/dx.bat" else="${android
tools}/dx" >
        <os family="windows"/>
    </condition>

    <property name="dx" value="${android-tools}/dx" />

    <property name="zip" value="zip" />
    <property name="android-jar" value="${sdk-folder}/android.jar" />

    <!-- Rules -->

    <!-- Create the output directories if they don't exist yet. -->
    <target name="dirs">
        <mkdir dir="${outdir}" />
        <mkdir dir="${outdir-classes}" />
    </target>

    <!-- Generate the R.java file for this project's resources. -->
    <target name="resource-src" depends="dirs">
        <echo>Generating R.java...</echo>
        <exec executable="${aapt}" failonerror="true">
            <arg value="compile" />
            <arg value="-m" />
            <arg value="-J" />
            <arg value="${outdir-r}" />
            <arg value="-M" />
            <arg value="AndroidManifest.xml" />
```

```
        <arg value="-S" />
        <arg value="${resource-dir}" />
        <arg value="-I" />
        <arg value="${android-jar}" />
    </exec>
</target>

<!-- Generate java classes from .aidl files. -->
<target name="aidl" depends="dirs">
    <apply executable="${aidl}" failonerror="true">
        <fileset dir="${srcdir}">
            <include name="**/*.aidl"/>
        </fileset>
    </apply>
</target>

<!-- Compile this project's .java files into .class files. -->
<target name="compile" depends="dirs, resource-src, aidl">
    <javac encoding="ascii" target="1.5" debug="true" extdirs=""
            srcdir="."
            destdir="${outdir-classes}"
            bootclasspath="${android-jar}" />
</target>

<!-- Convert this project's .class files into .dex files. -->
<target name="package-dex" depends="dex, package-res">
    <echo>Packaging dex...</echo>
    <exec executable="${zip}" failonerror="true">
        <!--<arg value="-Xmx384M" />-->
        <!-- Move Xmx parameter to dx.bat - FIX - BLOCK END -->
        <arg value="--dex" />
        <arg value="--output=${intermediate-dex}" />
        <arg value="--locals=full" />
        <arg value="--positions=lines" />
        <arg path="${outdir-classes}" />
    </exec>
</target>

<!-- Put the project's resources into the output package file. -->
<target name="package-res-and-assets">
    <echo>Packaging resources and assets...</echo>
    <exec executable="${aapt}" failonerror="true">
        <arg value="package" />
        <arg value="-f" />
        <arg value="-c" />
        <arg value="-M" />
        <arg value="AndroidManifest.xml" />
        <arg value="-S" />
        <arg value="${resource-dir}" />
```

```
            <arg value="-A" />
            <arg value="${asset-dir}" />
            <arg value="-I" />
            <arg value="${android-jar}" />
            <arg value="${out-package}" />
        </exec>
    </target>

    <!-- Same as package-res-and-assets, but without "-A ${asset-dir}" -->
    <target name="package-res-no-assets">
        <echo>Packaging resources...</echo>
        <exec executable="${aapt}" failonerror="true">
            <arg value="package" />
            <arg value="-f" />
            <arg value="-c" />
            <arg value="-M" />
            <arg value="AndroidManifest.xml" />
            <arg value="-S" />
            <arg value="${resource-dir}" />
            <!-- No assets directory -->
            <arg value="-I" />
            <arg value="${android-jar}" />
            <arg value="${out-package}" />
        </exec>
    </target>
    <!-- Invoke the proper target depending on whether or not
         an assets directory is present. -->
    <!-- TODO: find a nicer way to include the "-A ${asset-dir}" argument
         only when the assets dir exists. -->
    <target name="package-res">
        <available file="${asset-dir}" type="dir"
                property="res-target" value="and-assets" />
        <property name="res-target" value="no-assets" />
        <antcall target="package-res-${res-target}" />
    </target>

    <!-- Put the project's .class files into the output package file. -->
    <target name="package-java" depends="compile, package-res">
        <echo>Packaging java...</echo>
        <jar destfile="${out-package}"
                basedir="${outdir-classes}"
                update="true" />
    </target>
    <!-- Put the project's .dex files into the output package file. -->
    <target name="package-dex" depends="dex, package-res">
        <echo>Packaging dex...</echo>
        <exec executable="${zip}" failonerror="true">
            <arg value="-qj" />
            <arg value="${out-package}" />
            <arg value="${intermediate-dex}" />
        </exec>
```

```
    </target>
    <!-- Create the package file for this project from the sources. -->
    <target name="package" depends="package-dex" />
</project>
```

After modifying build.xml, you can then try to run **ant** again.

Installing Your Application with adb

The first step is to start your Emulator. In the Android /tools folder, find the emulator.exe file and execute it. This starts your Android server. That is, starting the Emulator also starts a virtual cell phone on your PC, as shown next. You can then use different tools to interact with the server, to do such things as install applications and call a shell environment.

To install your command-line application on your Android server, you need to use adb. adb is your connection to the Android server, which is started with your Emulator. adb contains many useful functions that allow you to interact with your Android server; one of these enables you to install applications.

Table 6-1 lists and describes the commands that adb accepts.

Command	Description
install <path>	Installs applications to the server
pull <remote file> <local file>	Pulls a remote file off the server
push <local file> <remote file>	Pushes a local file to the server
shell	Opens a shell environment on the sever
forward <local port> <remote port>	Forwards traffic from one port to another (to and from the server)
start-server	Starts the server
kill-server	Stops the server
ppp <tty> <params>	Uses a PPP connection over USB
devices	Lists the available emulators
help	Lists the adb commands
version	Displays the adb version

Table 6-1 adb Commands

To copy your application to the server, open a Windows command prompt and navigate to the directory of your build.xml file. The command syntax for adb is as follows:

```
adb install <apk path>
```

If the application installs to the phone properly, you will just get the package size as feedback in return from the command, as shown next.

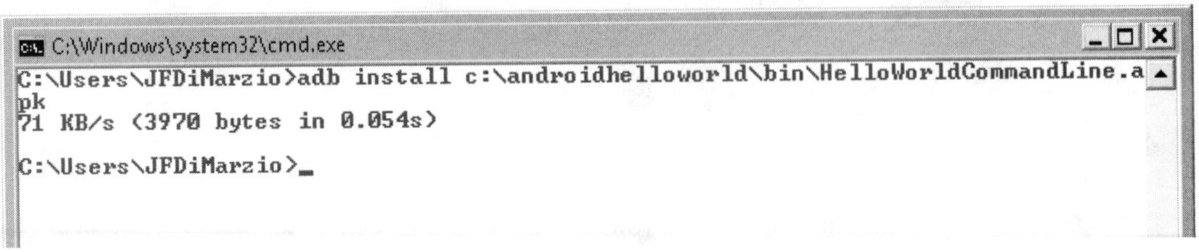

```
C:\Windows\system32\cmd.exe
C:\Users\JFDiMarzio>adb install c:\androidhelloworld\bin\HelloWorldCommandLine.a
pk
71 KB/s (3970 bytes in 0.054s)

C:\Users\JFDiMarzio>_
```

Switching over to your running Android Emulator, you should now see the application installed on your phone.

What to Do if Running the Application Produces an Error The first time I ran this application, after using the new build.xml file, I received an error on the Android Emulator. Shown in the following illustration, the error pointed to a missing class.

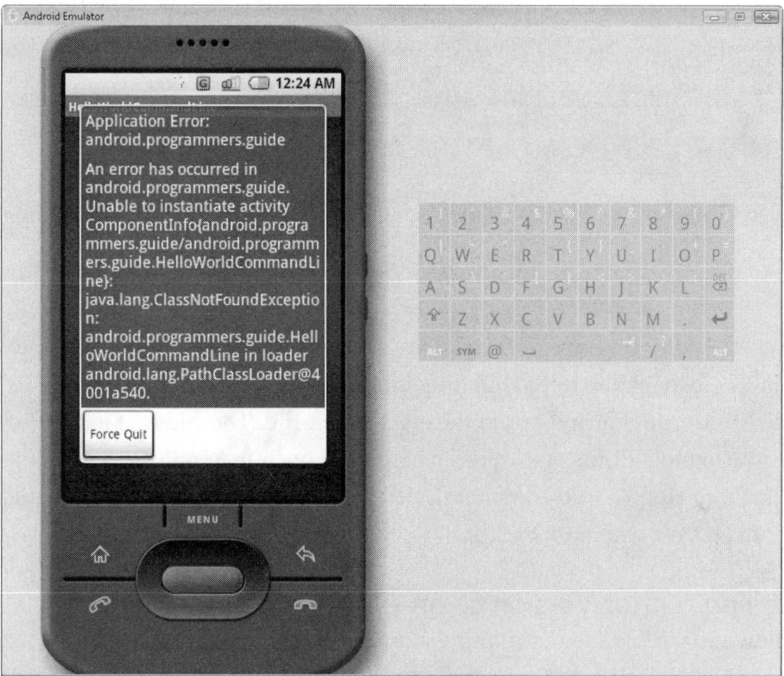

NOTE

While you may or may not encounter this exact error, depending on what release of the Android SDK is available when this book is published, you should follow the troubleshooting steps presented here, because they will help you in later projects.

This error seems to point to the fact that my classes, somehow, are missing from the HelloWorldCommandLine.apk file. I can easily remedy this issue without using any of the Android SDK command-line tools.

As it turns out, .apk files are just .zip files. That is, you can easily open them with a .zip file decompressor and view the files within them. The following illustration shows what the inside of HelloWorldCommandLine.apk looks like using the WinRAR decompressor.

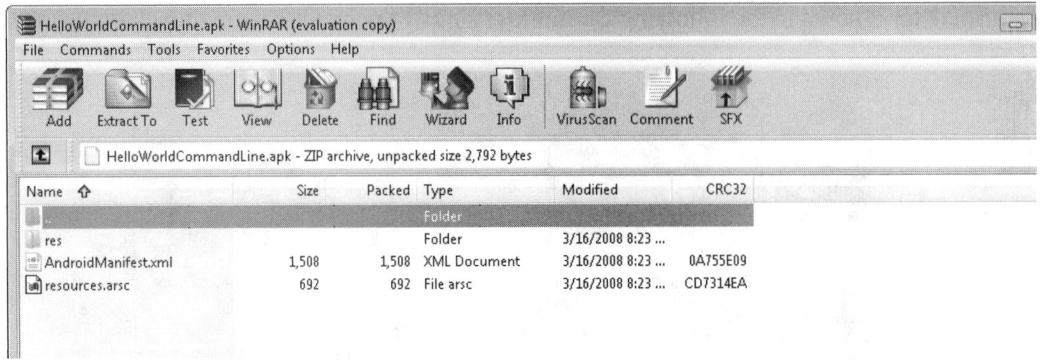

What is missing from the file is classes.dex. This is the compiled Dalvik executable of my classes. Navigating to the bin directory of my Android project, I can see that ANT did successfully compile and create the classes.dex file. The file was just left out of the HelloWorldCommandLine.apk. With the .apk file open in WinRAR, I can drag-and-drop classes.dex into HelloWorldCommandLine.apk. After classes.dex has been added to my .apk file, I can save and close the file.

Uninstalling a Prior Version of an Activity

Before you add the file to your running Android server, you are going to uninstall the prior version of HelloWorldCommandLine. Uninstalling a prior version of an application before you install another is not required. However, to get a good look at how to interact with the Android server, go ahead and uninstall your previous version before proceeding.

With your Android Emulator open, return to your command prompt environment and run the command **adb shell**, which opens the shell environment of the Android server. If you are successful, your command prompt should turn from a > to a #.

You now have an open shell into your Android server. There are a multitude of functions you can run from this point, but for now focus on one: removing the old HelloWorldCommandLine.apk file.

TIP

Keep in mind that Android is an operating environment. The commands that you can use within the shell are standard POSIX commands.

On the Android server, user-installed applications are kept in the /data/app directory. Using **cd**, navigate to the app directory, as shown in the following illustration.

```
Command Prompt - adb shell
Microsoft Windows [Version 6.0.6000]
Copyright (c) 2006 Microsoft Corporation.  All rights reserved.

C:\Users\JFDiMarzio>adb shell
# cd data
cd data
# cd app
cd app
# _
```

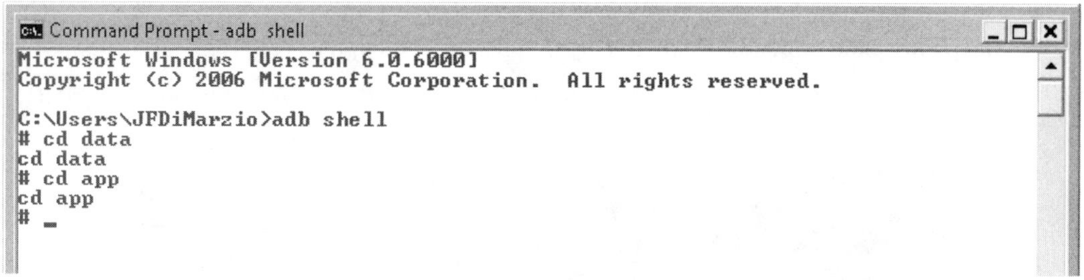

Run the command **ls** to list all the files in this directory. You should see a file named HelloWorldCommandLine.apk. This file represents the installation of your Activity.

Now that you have located the application on the Android server, you can remove it. Use the command syntax **rm HelloWorldCommandLine.apk** to remove the application. The following illustration shows that the **rm** command, if run correctly, gives no feedback. A subsequent use of **ls** shows that the file has been removed.

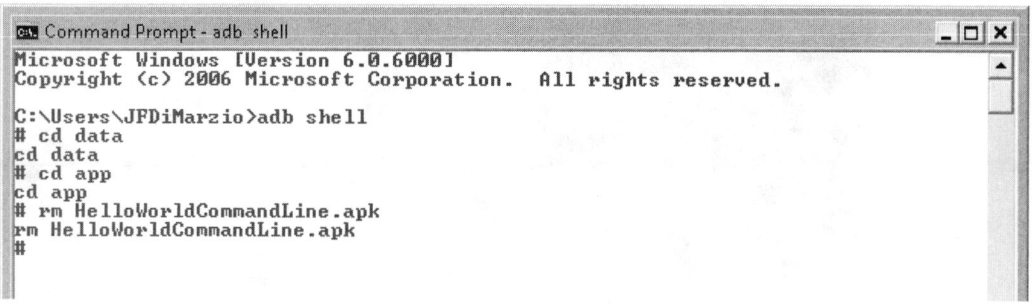

```
Command Prompt - adb shell
Microsoft Windows [Version 6.0.6000]
Copyright (c) 2006 Microsoft Corporation.  All rights reserved.

C:\Users\JFDiMarzio>adb shell
# cd data
cd data
# cd app
cd app
# rm HelloWorldCommandLine.apk
rm HelloWorldCommandLine.apk
#
```

CAUTION

Because you are technically logged into a Linux server via a shell, all the commands you run in the shell are case sensitive.

With the application removed, type **exit** to leave the shell and return to your command prompt.

Reinstalling and Launching the Application

You can now reinstall the application using **adb install**:

```
adb install HelloWorldCommandLine.apk
```

Once the application is installed back to the server, switch to your Emulator. Launch the application from your Emulator. It should work perfectly, as shown in the following illustration.

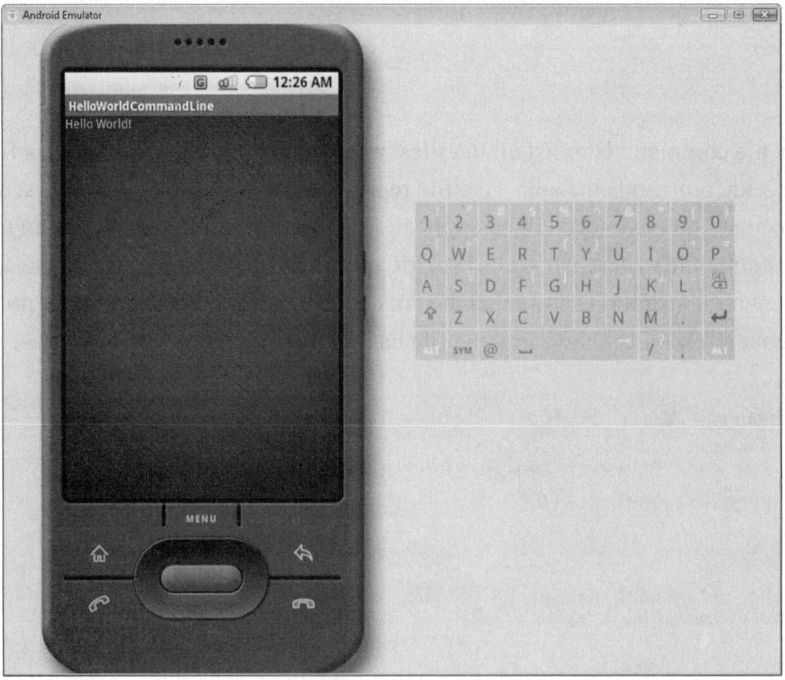

Now that we have covered the process for creating and editing files on Windows, let's take a look at it on Linux. Even if you are a die-hard Windows user, you may want to pay attention to the following section. I have found that there are definite advantages to programming with open source tools.

Hello World! on Linux

Many programmers, especially those who are interested in open source software, use Linux as their platform of choice. Google and the Open Handset Alliance have made an Android SDK just for these programmers. The SDK is actually the same SDK (because Java is portable), but the tools are created specifically to run on Linux.

When I started writing this book, I was using an older version of Red Hat Linux (Red Hat 9) as my Linux platform. I downloaded and installed Eclipse and the Android SDK. However, it quickly became apparent that there are limitations to the version of Linux that you can safely run Android on. As a minimum, you have to have a version of Linux that supports libstdc++.so.6.

The Android documentation lists Ubuntu Dapper Drake as a tested version of Linux. If you have not yet made a decision as to which version you want to use, you can feel safe with that version. Unfortunately, with the hardware that I am running, I had a problem installing the latest version of Ubuntu. So I decided to move away from what was recommended and try something new.

When I made the decision to drop Red Hat for another distribution of Linux, I decided to try Fedora 8. The remainder of this book uses Linux examples from Fedora 8; however, they should work without an issue on the distribution of your choice.

CAUTION

If you choose to use Fedora 8, it comes packaged with a custom version of Eclipse called Fedora Eclipse. If you attempt to install the Android plugin (using the steps outlined earlier in this book) for Fedora Linux, it will throw an error stating that the plugin org.eclipse.wst.sse.u is required. You can address this in either of two ways: download the latest version of Eclipse for Linux, or use Fedora's automatic update program, which will download an update to Fedora Eclipse that will bring it up to date with the latest version of Eclipse. You can then use this version of Eclipse with the Android SDK.

Configuring the PATH Statement

The first step is to configure your PATH statement. The path is the list of directories within which the operating system will look when trying to find a command that is being run. To see what your path is currently configured to, run the following from a terminal:

```
echo $PATH
```

You will get back something that resembles the PATH statement in the following illustration.

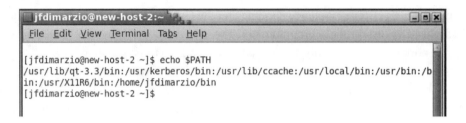

```
[jfdimarzio@new-host-2 ~]$ echo $PATH
/usr/lib/qt-3.3/bin:/usr/kerberos/bin:/usr/lib/ccache:/usr/local/bin:/usr/bin:/b
in:/usr/X11R6/bin:/home/jfdimarzio/bin
[jfdimarzio@new-host-2 ~]$
```

Use the **export** command to add Android to the PATH statement (see the next illustration):

```
export  PATH=$PATH:<android path>
```

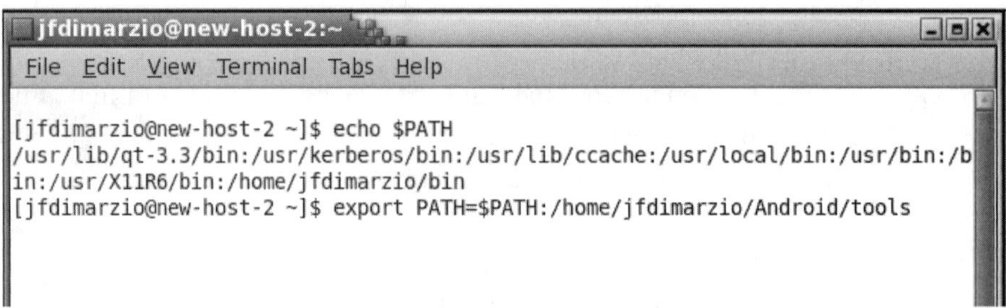

Editing the PATH statement like this in Linux will change the PATH statement only for the current terminal session. To make your PATH statement change permanent, you must edit .bash_profile. Use vi to edit .bash_profile, as shown in the following illustration.

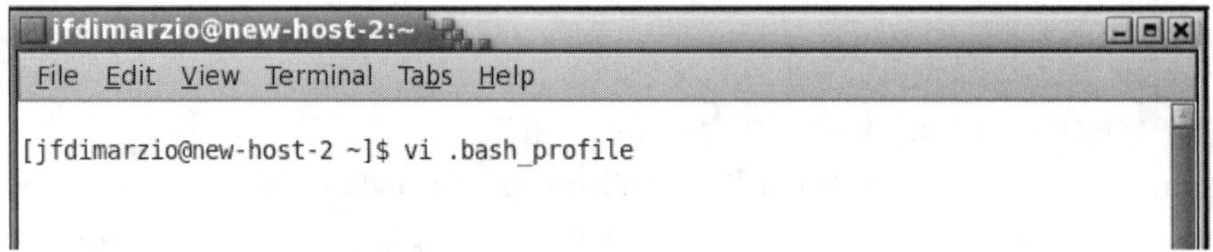

With .bash_profile open in the vi editor, it should look something like the next illustration. As you can see the PATH statement is clearly visible. Use the command

:i to put vi in insert mode, and then add Android to the PATH statement. Then press the ESC key, use the command **:w** to write the file, and then use **:q** to quit.

```
# .bash_profile

# Get the aliases and functions
if [ -f ~/.bashrc ]; then
        . ~/.bashrc
fi

# User specific environment and startup programs

PATH=$PATH:$HOME/bin:$HOME/Android/tools

export PATH
~
~
~
~
~
~
~
~
~
~
~
".bash_profile" 12L, 196C                                10,1          All
```

The Linux version of the Android SDK comes with a Python script, activityCreator.py, that is used to create your initial projects. When running the Python script, an output directory is created for your project. However, I like to create this directory manually to make sure it is created where I need it to be. Use **mkdir** to create a directory for your project (see the following illustration).

```
[jfdimarzio@new-host-2 ~]$ mkdir androidHelloWorld
[jfdimarzio@new-host-2 ~]$
```

After you create the project directory, you can run the activityCreator.py Python script. The syntax for the script is very close to that of the Windows .bat file:

```
activityCreator.py --out <output directory> package.activityName
```

Use the activityCreator.py script to set up your project. Take a look at the following illustration to see the output from the activityCreator.py script.

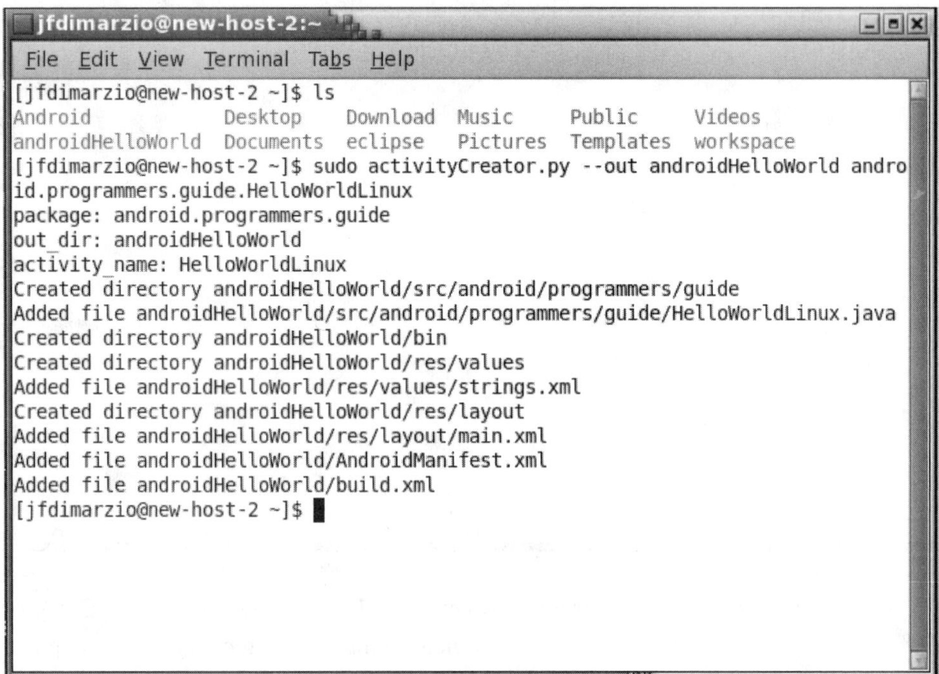

jfdimarzio@new-host-2:~
File Edit View Terminal Tabs Help
[jfdimarzio@new-host-2 ~]$ ls
Android Desktop Download Music Public Videos
androidHelloWorld Documents eclipse Pictures Templates workspace
[jfdimarzio@new-host-2 ~]$ sudo activityCreator.py --out androidHelloWorld andro
id.programmers.guide.HelloWorldLinux
package: android.programmers.guide
out_dir: androidHelloWorld
activity_name: HelloWorldLinux
Created directory androidHelloWorld/src/android/programmers/guide
Added file androidHelloWorld/src/android/programmers/guide/HelloWorldLinux.java
Created directory androidHelloWorld/bin
Created directory androidHelloWorld/res/values
Added file androidHelloWorld/res/values/strings.xml
Created directory androidHelloWorld/res/layout
Added file androidHelloWorld/res/layout/main.xml
Added file androidHelloWorld/AndroidManifest.xml
Added file androidHelloWorld/build.xml
[jfdimarzio@new-host-2 ~]$

TIP

Notice that the activityCreator.py command is prefixed by **sudo**. The **sudo** command is used to emulate the permissions of another user (in this case, root) if you do not have sufficient permissions to run the requested command. On my installation of Fedora, my user account does not have the rights to interact with certain directories the way root does.

With the project created, edit HelloWorldLinux.java to add the TextView. You can choose to edit the .java file a number of ways in Linux. You can use vi once again, or you can use a standard text editor as shown in the following illustration.

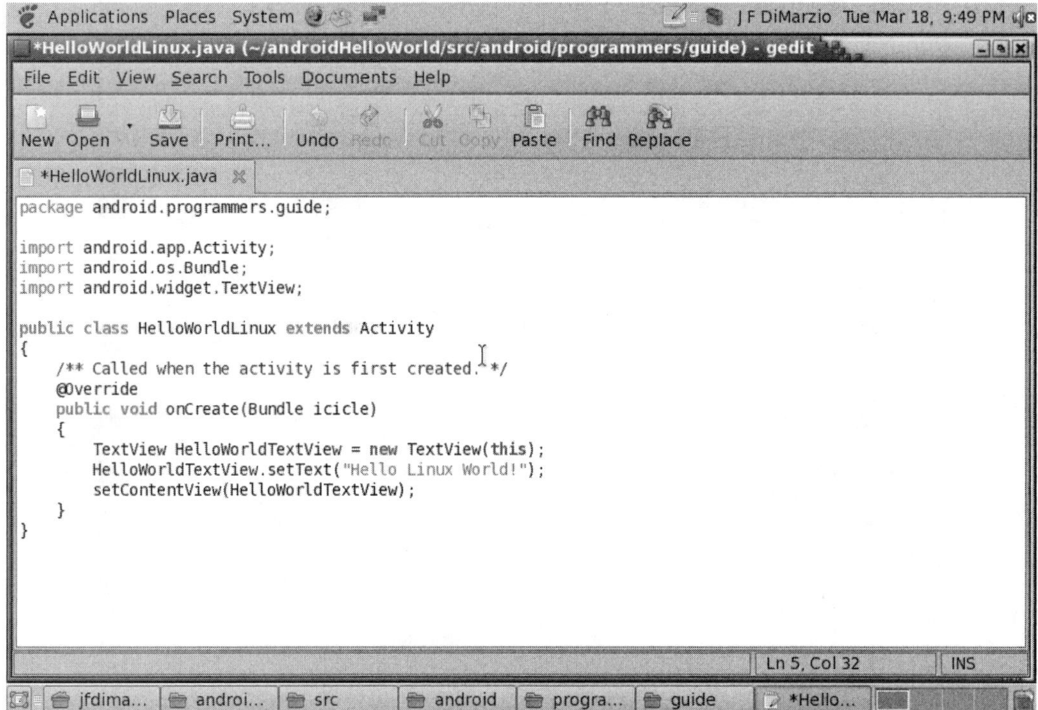

Finally, remove the defined TextView from main.xml. These two small changes are all you need to now compile your Linux version of the Hello World! application.

To compile the application, use ANT (which is what was used in the Windows environment earlier in the chapter). Apache ANT should be preinstalled in your Linux distribution, especially if you are using Fedora 8. If you are not using Fedora 8, you need to download, install, and set the path for the Linux version of Apache ANT.

When you run **ant**, you should see an output like that shown in the following illustration.

```
jfdimarzio@new-host-2:~/androidHelloWorld
File Edit View Terminal Tabs Help
[jfdimarzio@new-host-2 ~]$ cd androidHelloWorld
[jfdimarzio@new-host-2 androidHelloWorld]$ ant
Buildfile: build.xml

dirs:

resource-src:
     [echo] Generating R.java...
     [exec]     (skipping backup file 'res/layout/main.xml~')

aidl:

compile:
    [javac] Compiling 2 source files to /home/jfdimarzio/androidHelloWorld/bin/c
lasses

dex:

package-res:

package-res-no-assets:
     [echo] Packaging resources...
     [exec]     (skipping backup file 'res/layout/main.xml~')

package-dex:
     [echo] Packaging dex...

package:

BUILD SUCCESSFUL
Total time: 5 seconds
[jfdimarzio@new-host-2 androidHelloWorld]$
```

Finally, you need to start up your Android Emulator and install your application. With the Emulator started and running, execute the following command:

```
adb install HelloWorldLinux.apk
```

This installs the application to the Linux Android server. If the command runs successfully, you should be able to run your Activity in the Android Emulator.

The next chapter explores how to use the Android SDK to react to phone events.

Try This Create an Image-Based Hello World! in the CLI

Using the command-line tools covered in this chapter, re-create the image-based Hello World! project from Chapter 5. When you are creating this project, keep the following items in mind:

- Place the image in the res folder.
- Check if any tools are needed to create an R.java file with a handle into the image.
- Compile the project using ANT.
- Use the command **adb install** to push the application to your emulator.

Ask the Expert

Q: Is one operating system better than any other when programming for Android?

A: After using several operating systems with Android, I have not noticed any one operating system having a clear and distinct advantage over another. It is really just a matter of personal preference. However, as often happens, you may see more "tools"—of the unofficial sort—be released on the Linux platform. Because both Linux and Android are open source, more open source developers will be apt to create tools for other open source platforms. This symbiosis may even end up benefiting Android more than it benefits Linux.

Q: Are there other commands that can be run from within the adb shell environment?

A: Yes. For example, one interesting command is the **service** command, which can be used to check on the status of a process, such as:

```
service check phone
```

Assuming the phone is running, you should get back

```
Service phone: found
```

Another use of the **service** command is to place calls. With the Emulator running, type the following command and check the results on the Emulator interface:

```
service call phone 2 s16 "15555551212"
```

Chapter 7

Using Intents and the Phone Dialer

Key Skills & Concepts

- Using Intents
- Creating code that interacts with the phone hardware
- Learning the difference between dialing and calling

The chapters up to this point have introduced you to the basics of Android programming. You have examined the outline of an Android application and installed your first applications to the Android server. You have learned how to use Views and setContentView( ), as well as how to create a UI in XML. These skills have helped you to create a static application. What you have not done yet is use the application interface to interact with the hardware that the platform was created for—the cell phone.

You should not lose sight of the fact that the platform for which Android was created is, in essence, still a cell phone. The underlying hardware for the devices that Android will run on is designed for the purpose of person-to-person communication. If you strip away all the bells and whistles that the Android SDK is capable of adding to the cell phone, it must still be able to send and receive phone calls. For this reason, this chapter focuses on the code that enables you to interact with the phone hardware.

By the end of this chapter, you should have the skills needed to interact with some of the basic functions of the phone. You will be able to work with the dialer to send and receive calls. These tools and skills will be your keys to creating useful applications on this flexible platform.

You are reading this book because you intend to design applications that run on a cell phone, so it stands to reason that you should learn how Android allows for interaction with the phone hardware—in particular, the process that enables the phone to send and receive calls.

When we think of a cell phone, a few basic functions come to mind. The first, and most obvious, of which is the ability to send and receive phone calls. This is inarguably the quintessential function of a cell phone. There are a few peripheral features that make the cell phone easier to use, such as the ability to keep and manage contacts and the ability to store and review missed calls. As you'll read in this chapter, you can access and manipulate the code for all of these functions.

The first phone function that you will look at in this chapter is the sending of calls. You will create an application, using an Intent, that controls the phone dialer and causes it to place a call to a number. As the chapter progresses, you will expand on this application and add some bells and whistles to it.

NOTE

On the Android platform, there is a difference between the actions of dialing and calling. When you *dial* a number, you enter the digits into the keypad (or programmatically), but no call is actually placed. That is, dialing does not encompass pressing the Send button. When you *call* a number, you send a signal from your handset. That is, after you enter the number into the dialer, you press the Send button—either physically or programmatically. You need to know the difference between the two actions to understand the scope of the applications you will create in this section.

What Are Intents?

Before you can begin to interact with the phone dialer, you need to understand the type of code that you will use to do the job. Android uses *Intents* to do specific jobs within applications. Once you master the use of Intents, a whole new world of application development will be open to you. This section defines what an Intent is and how it is used.

An Intent is Android's method for relaying certain information from one Activity to another. An Intent, in simpler terms, expresses to Android your intent to do something. You can think of an Intent as a message passed between Activities. For example, assume that you have an Activity that needs to open a web browser and display a page on your Android device. Your Activity would send an "intent to open *x* page in the web browser," known as a WEB_SEARCH_ACTION Intent, to the Android Intent Resolver. The Intent Resolver parses through a list of Activities and chooses the one that would best match your Intent; in this case, the Web Browser Activity. The Intent Resolver then passes your page to the web browser and starts the Web Browser Activity.

Intents are broken up into two main categories:

- **Activity Action Intents** Intents used to call Activities outside of your application. Only one Activity can handle the Intent. For example, for a web browser, you need to open the Web Browser Activity to display a page.
- **Broadcast Intents** Intents that are sent out for multiple Activities to handle. An example of a Broadcast Intent would be a message sent out by Android about the current battery level. Any Activity can process this Intent and react accordingly—for example, cancel an Activity if the battery level is below a certain point.

Table 7-1 lists and describes the current Activity Action Intents that are available to you. As you'll notice, in most cases, the name of the Intent does a good job of describing what that Intent does.

Activity Action Intent	Message
ADD_SHORTCUT_ACTION	Add a function shortcut to the Android Home Screen
ALL_APPS_ACTION	List all the applications available on the device
ANSWER_ACTION	Answer an incoming call
BUG_REPORT_ACTION	Open the Bug Reporting Activity
CALL_ACTION	Place a call to supplied location
DELETE_ACTION	Delete the specified data
DIAL_ACTION	Open the Dial Activity and dial the specified number
EDIT_ACTION	Provide editable access to the supplied data
EMERGENCY_DIAL_ACTION	Dial an emergency number
FACTORY_TEST_ACTION	Retrieve factory test settings
GET_CONTENT_ACTION	Select and return specified data
INSERT_ACTION	Insert an empty item
MAIN_ACTION	Establish the Activity start point
PICK_ACTION	Pick an item and return the selection
PICK_ACTIVITY_ACTION	Pick a given Activity (returns a class)
RUN_ACTION	Execute the given data
SEARCH_ACTION	Launch a search on the system
SEND_ACTION	Send data without specifying the recipient
SENDTO_ACTION	Send data to the recipient specified
SETTINGS_ACTION	Launch System Settings
SYNC_ACTION	Sync phone data with external source
VIEW_ACTION (DEFAULT_ACTION)	Open a View
WALLPAPER_SETTINGS_ACTION	Show settings for modifying the Android Wallpaper
WEB_SEARCH_ACTION	Open Google Search, or another web page if specified

Table 7-1 Activity Action Intents

NOTE

For the applications in this chapter, you will use two of the Intents listed in Table 7-1: CALL_ACTION and DIAL_ACTION. These Intents give you access to the phone's dialing and calling capabilities.

Table 7-2 lists and describes the current Broadcast Intents that are available. Refer to this list when you need to establish a receiver for a specific Intent.

Broadcast Intent	Message
CALL_FORWARDING_STATE_CHANGED_ACTION	The phone's call forwarding state has changed
CAMERA_BUTTON_ACTION	The camera button has been pressed
CONFIGURATION_CHANGED_ACTION	The device's configuration has changed
DATA_ACTIVITY_STATE_CHANGED_ACTION	The device's data activity state has changed
DATA_CONNECTION_STATE_CHANGED_ACTION	There has been a change in the data connection state
DATE_CHANGED_ACTION	The phone's system date has changed
FOTA_CANCEL_ACTION	Cancel pending system update downloads
FOTA_INSTALL_ACTION	An update has been downloaded and must be installed immediately (sent by the system)
FOTA_READY_ACTION	An update has been downloaded and can be installed—but does not need to be installed immediately (sent by the system)
FOTA_RESTART_ACTION	Restart a system update download
FOTA_UPDATE_ACTION	Begin the download of a system update
GTALK_SERVICES_CONNECTED_ACTION	Sent when a GTALK session has been successfully established
GTALK_SERVICES_DISCONNECTED_ACTION	Sent when a GTALK session has been disconnected
MEDIA_BAD_REMOVAL_ACTION	Sent when an SD Memory Card was removed but unsuccessfully unmounted from the system
MEDIA_BUTTON_ACTION	Sent when the media button has been pressed

Table 7-2 Broadcast Intents

Broadcast Intent	Message
MEDIA_EJECT_ACTION	Sent when the eject action has been initiated on an SD Memory Card
MEDIA_MOUNTED_ACTION	Sent when an SD Memory Card was successfully mounted to the system
MEDIA_REMOVED_ACTION	Sent when an SD memory card was detected as having been removed
MEDIA_SCANNER_FINISHED_ACTION	Sent when the scanner has finished
MEDIA_SHARED_STARTED_ACTION	Sent when the scanner has begun
MEDIA_UNMOUNTED_ACTION	Sent when an SD memory card has been detected but has not been mounted
MESSAGE_WAITING_STATE_CHANGED	The "message waiting" state on the phone has changed
NETWORK_TICKLE_RECEIVED_ACTION	A new device network notification has been received
PACKAGE_ADDED_ACTION	Sent when a new package has been installed on the device
PACKAGE_CHANGE_ACTION	Sent when an existing package has been modified
PACKAGE_INSTALL_ACTION	A package can be downloaded and installed
PACKAGE_REMOVED_ACTION	A package has been removed
PHONE_INTERFACE_ADDED_ACTION	The device's phone interface has been created
PHONE_STATE_CHANGED_ACTION	The device's phone state has changed
PROVIDER_CHANGED_ACTION	The device has received a notification from a provider
PROVISIONING_CHECK_ACTION	Check for the latest settings from the provisioning service
SCREEN_OFF_ACTION	The screen has been shut off (sent by the device)
SCREEN_ON_ACTION	The screen has been turned on (sent by the device)
SERVICE_STATE_CHANGED_ACTION	The service state has changed
SIGNAL_STRENGTH_CHANGED_ACTION	The signal strength has changed

Table 7-2 Broadcast Intents *(continued)*

Broadcast Intent	Message
SIM_STATE_CHANGED_ACTION	The state of the SIM card has changed
TIME_CHANGED_ACTION	The device's time was set
TIME_TICK_ACTION	The current time has changed
TIMEZONE_CHANGED_ACTION	The device's timezone has changed
UMS_CONNECTED_ACTION	The device has connected via USB
UMS_DISCONNECTED_ACTION	The device has been disconnected from its USB host
WALLPAPER_CHANGED_ACTION	The device's wallpaper has been changed

Table 7-2 Broadcast Intents *(continued)*

NOTE

Some of these Broadcast Intents are sent out quite often, such as TIME_TICK_ACTION and SIGNAL_STRENGTH_CHANGED_ACTION. Be careful how you use them. You should try not to receive such broadcasts if at all possible.

The Intent is only one-third of the picture. An Intent is really just that, an intent to do something; an Intent cannot actually do anything by itself. You need Intent Filters and Intent Receivers to listen for, and interpret, the Intents.

An Intent Receiver is like the mailbox of an Activity. The Intent Receiver is used to allow an Activity to receive the specified Intent. Using the previous web browser example, the Web Browser Activity is set up to receive web browser Intents. A system like this allows unrelated Activities to ignore Intents that they would not be able to process. It also allows Activities that need the assistance of another Activity to utilize that Activity without needing to know how to call it.

With Intents and Intent Receivers, one Activity can send out an Intent and another can receive it. However, there needs to be something that governs the type of information that can be sent between the two Activities. This is where Intent Filters come in.

Intent Filters are used by Activities to describe the types of Intents they want to receive. More importantly, they outline the type of data that should be passed with the Intent. Therefore, in our example scenario, we want the web browser to open a web page. The Intent Filter would state that the data passed with the WEB_SEARCH_ACTION Intent should be in the form of a URL.

In the next section, you will begin to use Intents to open and utilize the phone's dialer.

Using the Dialer

Now that you know what an Intent is, it is time to see one in action. This section shows you how to use the DIAL_ACTION Intent to open the phone dialer. You will pass a telephone number with your Intent. If your application works correctly, you should see displayed in the dialer the number you pass with your Intent.

The first step is to create a new project for this Activity (see Chapter 5 for instructions). Name the project **AndroidPhoneDialer**. The following illustration shows the New Android Project wizard for this project.

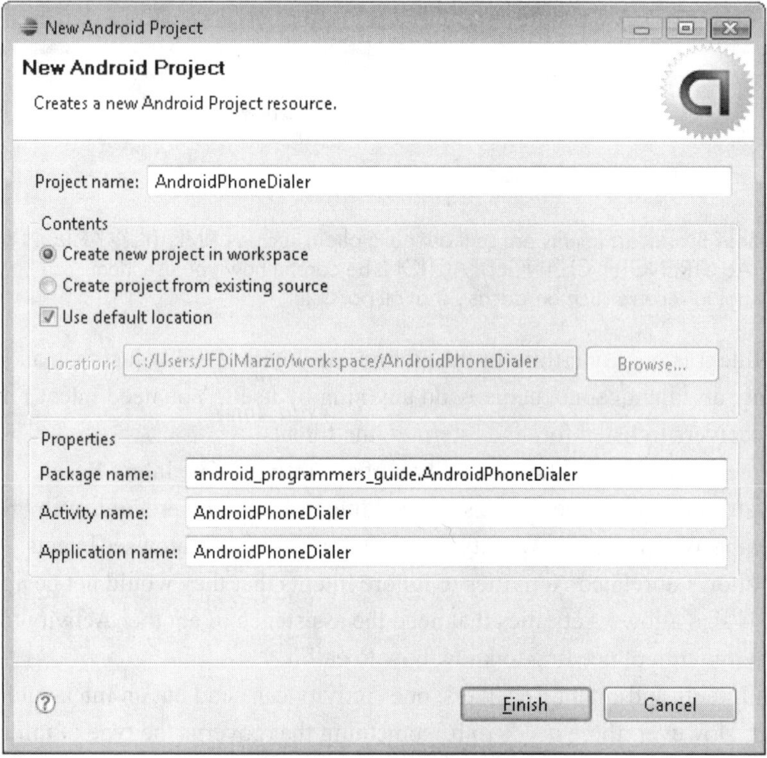

With your new application open in Eclipse, the first order of business is to remove the TextView from main.xml that contains that Hello World statement. The main.xml file should look like this after you remove the TextView:

```
<?xml version="1.0" encoding="utf-8"?>
<LinearLayout xmlns:android=http://schemas.android.com/apk/res/android
```

```
    android:orientation="vertical"
    android:layout_width="fill_parent"
    android:layout_height="fill_parent"
    >
</LinearLayout>
```

You need to add two new packages to your project to utilize the DIAL_ACTION Intent, as follows. The first package allows you to set up Intents and the second allows you to parse URIs.

```
import android.content.Intent;
import android.net.Uri;
```

NOTE

There are several different Intent Filters on the DIAL_ACTION Intent that you can use. You are using the Filter that lets you pass a phone number as a URI.

The next step is to create your Intent. The syntax for creating an Intent is as follows:

```
Intent <intent_name> = new Intent(<Android_Intent>,<data>)
```

For your application, replace the first parameter, *<intent_name>*, with **DialIntent**. To get the value for the second parameter, *<Android_Intent>*, refer to the list of Activity Actions in Table 7-1. You'll find that, to call the dialer, you need to use the DIAL_ACTION Intent. To call the Intent properly, use the format **Intent.DIAL_ACTION**. The last parameter, *<data>*, is the phone number. The DIAL_ACTION Intent takes in data as a URI. Thus, you need to use Uri.parse to parse out your phone number. Using Uri.parse will ensure that the DIAL_ACTION Intent understands the number you are trying to dial. You pass Uri.parse a string that represents the phone number you want to dial, "tel:5551212" in this example.

The final call to create an Intent for your project should look like this:

```
        Intent DialIntent = new
Intent(Intent.DIAL_ACTION,Uri.parse("tel:5551212"));
```

TIP

You use the notation *tel:<phone_number>* to dial a specific phone number. You can also use *voicemail:* instead of *tel:* to dial the phone's voicemail shortcut.

With the Intent created, you now have to tell Android that you want the dialer to be launched in a new Activity. To do this, you use the setLaunchFlags( ) method of the Intent. You must pass setLaunchFlags( ) the appropriate parameter for launching. The following is a list of the possible launch flags that you can set:

NOTE

In some cases, more than one of the following flags may be set to accomplish the desired outcome.

- **NO_HISTORY_LAUNCH** Launches the Activity without recording it in the system's launch history
- **SINGLE_TOP_LAUNCH** Tells the system not to launch the Activity if it is already running
- **NEW_TASK_LAUNCH** Launches the Activity
- **MULTIPLE_TASK_LAUNCH** Launches the Activity even if it is already running
- **FORWARD_RESULT_LAUNCH** Allows the new Activity to receive results that would normally be forwarded to the existing Activity

In this case, you want to use Intent.NEW_TASK_LAUNCH, which simply lets you open a new instance of the dialer Activity:

```
DialIntent.setLaunchFlags(Intent.NEW_TASK_LAUNCH );
```

The last step to creating your dialer Intent is to actually launch the Activity. (More accurately, you are telling Android that you have an intent to launch the dialer as a New Task. It is ultimately up to Android to launch the Dialer Activity.) To tell Android that you want to start the dialer, you need to use startActivity( ):

```
startActivity(DialIntent);
```

Notice that you pass to startActivity( ) your Intent. The Intent is then passed to Android, and the action is resolved. The full code for AndroidPhoneDialer.java should look like this:

```
package android_programmers_guide.AndroidPhoneDialer;
import android.app.Activity;
import android.content.Intent;
import android.os.Bundle;
import android.net.Uri;

public class AndroidPhoneDialer extends Activity {
    /** Called when the Activity is first created. */
    @Override
    public void onCreate(Bundle icicle) {
        super.onCreate(icicle);
        setContentView(R.layout.main);
       /** Create our Intent to call the Dialer */
       /** Pass the Dialer the number 5551212   */
        Intent DialIntent = new
Intent(Intent.DIAL_ACTION,Uri.parse("tel:5551212"));
       /** Use NEW_TASK_LAUNCH to launch the Dialer Activity */
        DialIntent.setLaunchFlags(Intent.NEW_TASK_LAUNCH );
       /** Finally start the Activity          */
        startActivity(DialIntent);
    }
}
```

You should now compile AndroidPhoneDialer and run it on your Emulator. The process for compiling and running applications was covered in previous chapters, so you should be familiar with that process. Once you run your application, the Emulator should launch. After the lengthy boot process, your Activity will launch.

TIP

It is a good idea to keep the Emulator running, even after you are finished with your Activity and have returned to the code window. It is most people's instinct to close the Emulator window when they have finished testing their Activity. However, I have found that leaving the Emulator open helps with two major issues. The first is the amount of time it takes for the Emulator to start. By leaving the Emulator open, you avoid the lengthy load time. Second, I have noticed that there are times when I make minor changes to an Activity and they are not copied to the Emulator. Leaving the Emulator open seems to alleviate this issue as well. If you continue to have issues in the Emulator, remove the userdata-qemu.img file from your computer. This allows the Emulator to start up with a clean image.

If you have followed the code in the examples correctly, you should see the following:

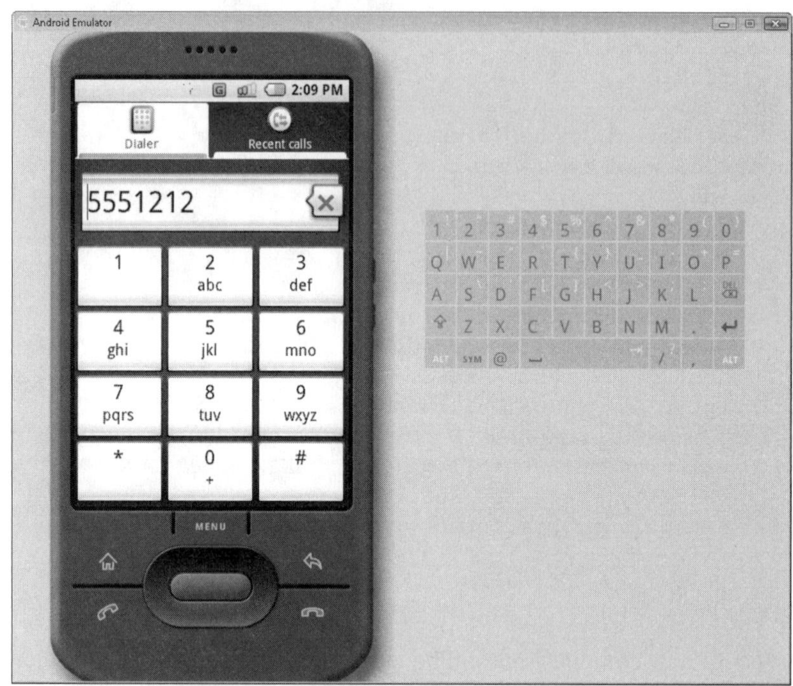

As you can see, you have now opened the phone's Dialer Activity. The Dialer is displaying the phone number that you passed to it, 5551212. Using the Emulator, click the Send button, and the phone should now call 555-1212—virtually of course.

Just displaying the Dialer Activity is useful only if you want to create an application that allows the user to edit the number before calling it, or even confirm that they really want to make the call. What should you do if you want your application to actually place the call for you? The answer is addressed next.

Placing a Call from Your Activity

In this section you will learn what Intent to add to your Activity when calling the dialer. You will also learn where to add your chosen Intent in the Activity's code. Further, you will learn how to parse the intended phone number as a URI.

You need to make a few changes to your code to move from the Dialer Activity to the Call Activity. In this section, you are going to edit your AndroidPhoneDialer Activity to place a call after opening the dialer.

Adding the Intent to Your Activity

You still need the Intent and Uri packages—shown here—so leave those in place at the header of your AndroidPhoneDialer.java file.

```
import android.content.Intent;
import android.net.Uri;
```

These packages will enable you to not only instantiate the Intent that you need, but also pass the needed telephone number data to the Intent (with the Uri package).

TIP

If you are flipping through the chapters out of order, and did not work on the project in the previous section, simply create a new project, name it AndroidPhoneDialer, and add the previous two packages to it. That will catch you up to speed.

Take a look now through the list of possible Activity Action Intents in Table 7-1, shown earlier in this chapter. True to its name, the Intent that you need in your Activity is CALL_ACTION. In much the same way that DIAL_ACTION opened the Android dialer, CALL_ACTION will launch the phone's calling process and initiate a call to the supplied number.

To create the Intent, use the same procedure as you did for the dialer, only this time call CALL_ACTION:

```
        Intent CallIntent = new
Intent(Intent.CALL_ACTION,Uri.parse("tel:5551212"));
```

Notice that you use Uri.parse to pass a correctly parsed telephone number to the Activity. The next step is to tell Android that you want to set this Activity to launch, and then launch it. This is accomplished using the following two lines of code.

```
        CallIntent.setLaunchFlags(Intent.NEW_TASK_LAUNCH );
        startActivity(CallIntent);
```

In the first line, you sent the launch flag to NEW_TASK_LAUNCH. This launches a new instance of the Call Activity. Finally, you tell Android to start the Activity using your Intent. When finished, your AndroidPhoneDialer.java file should look like this.

```
package android_programmers_guide.AndroidPhoneDialer;
import android.app.Activity;
```

```
import android.content.Intent;
import android.os.Bundle;
import android.net.Uri;

public class AndroidPhoneDialer extends Activity {
    /** Called when the Activity is first created. */
    @Override
    public void onCreate(Bundle icicle) {
        super.onCreate(icicle);
        setContentView(R.layout.main);
       /** Create our Intent to call the device's Call Activity */
       /** Pass the Call the number 5551212   */
        Intent CallIntent = new
Intent(Intent.CALL_ACTION,Uri.parse("tel:5551212"));
       /** Use NEW_TASK_LAUNCH to launch the Call Activity */
        CallIntent.setLaunchFlags(Intent.NEW_TASK_LAUNCH );
       /** Finally start the Activity           */
        startActivity(CallIntent);
    }
}
```

Compile the application now and observe the results; you should see something similar to the error in the following illustration.

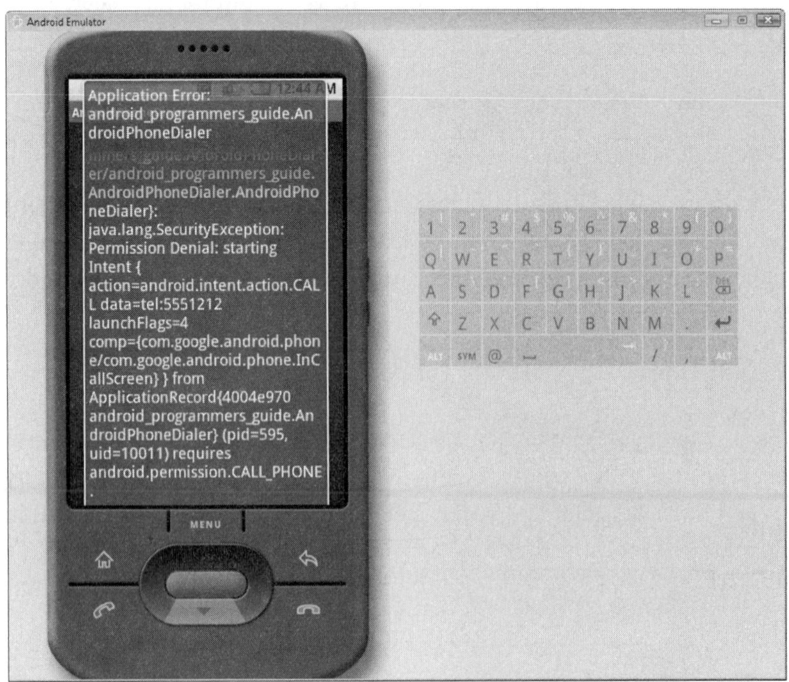

I purposely wanted you to see this error because it shows a side of Android that we have not explored yet. The text of the error should read as follows:

```
Application_Error:
...
Java.lang.SecurityException:
Permission Denial: starting Intent
...
```

Android marshals certain actions by requiring that permissions be granted to perform them, as covered next.

Editing Activity Permissions

Most Activity Action Intents fall into the category of requiring that the proper permission be set before Android will allow the action. As with many systems, Android just needs to make sure that only Activities with the correct credentials be allowed to perform actions with Activities that are outside of their base. Here are the available permissions:

- ACCESS_ASSISTED_GPS
- ACCESS_CELL_ID
- ACCESS_GPS
- ACCESS_LOCATION
- ACCESS_SURFACE_FLINGER
- ADD_SYSTEM_SERVICE
- BROADCAST_PACKAGE_REMOVED
- BROADCAST_STICKY
- CALL_PHONE
- CHANGE_COMPONENT_ENABLED_STATE
- DELETE_PACKAGES
- DUMP
- FOTA_UPDATE
- GET_TASKS
- INSTALL_PACKAGES
- INTERNAL_SYSTEM_WINDOW
- RAISED_THREAD_PRIORITY
- READ_CONTACTS
- READ_FRAME_BUFFER
- RECEIVE_BOOT_COMPLETED
- RECEIVE_SMS
- RECEIVE_WAP_PUSH
- RUN_INSTRUMENTATION
- SET_ACTIVITY_WATCHER
- SET_PREFERRED_APPLICATIONS
- SIGNAL_PERSISTENT_PROCESSES
- SYSTEM_ALERT_WINDOW
- WRITE_CONTACTS
- WRITE_SETTINGS

Compare this list of permissions with the list of Intents in Table 7-1. You should find that most of the Intents match up with a corresponding permission. The CALL_ACTION Intent is no exception. You need to assign your Activity the CALL_PHONE permission to be able to execute your Intent.

To assign your Activity the correct permission, you first need to know what permission you need to assign. The current example is using the Dialer Activity. Access to the Dialer Activity is governed by the CALL_PHONE permission. By assigning this permission to your Activity, Android will let your Intent launch the Dialer Activity.

How do you add permissions to the Activity? You need to edit the Activity's Manifest. If you are using Eclipse, double-click AndroidManifest.xml. This opens the Android Manifest Overview window, shown in the following illustration.

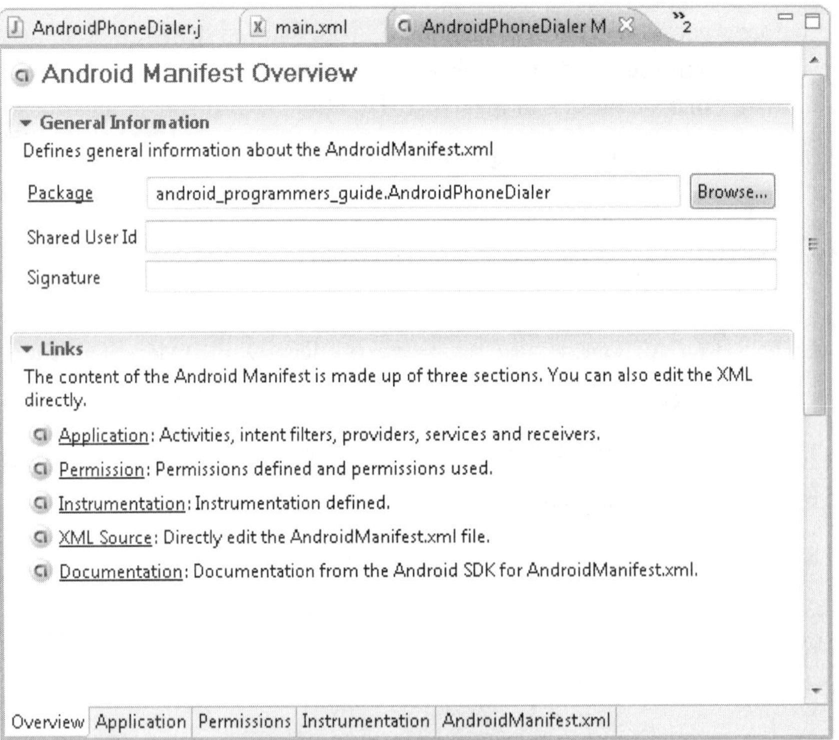

To edit the Activity's permissions, click the Permission link. This should take you to the Android Manifest Permissions window, shown in the following illustration.

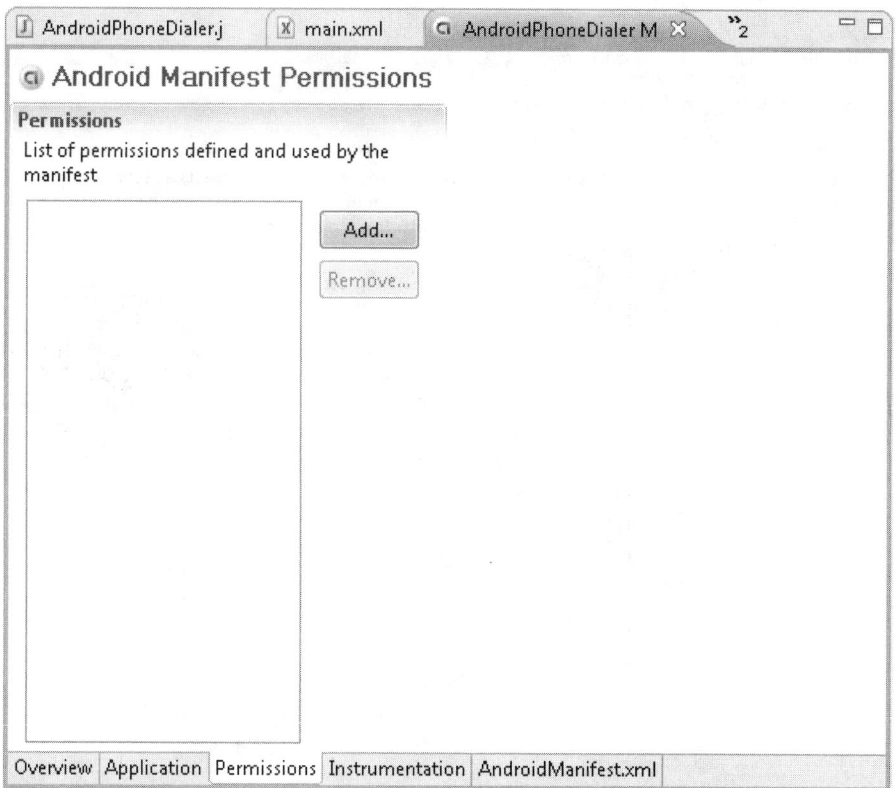

This window lists the permissions that are currently assigned to your Activity. Given that you are working in a new project, you do not have any assigned permissions. Therefore, click the Add button to begin the process. In the dialog box that opens, select Uses Permission and click OK.

Back in the Android Manifest Permissions window, in the Name drop-down list, select android.permission.CALL_PHONE, as shown next. This adds the CALL_PHONE permission to your Activity.

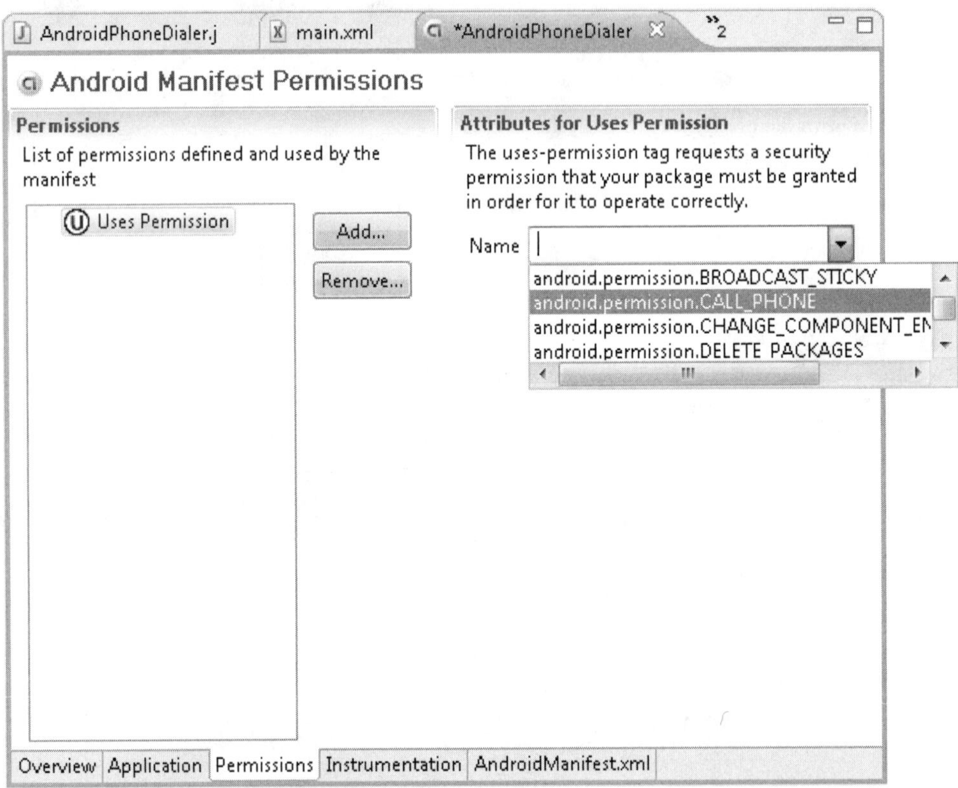

Now that you have added the CALL_PHONE permission, take a look at AndroidManifest.xml. It should look similar to the following:

```
<?xml version="1.0" encoding="utf-8"?>
<manifest xmlns:android=http://schemas.android.com/apk/res/android
    package="android_programmers_guide.AndroidPhoneDialer">
    <application android:icon="@drawable/icon">
        <activity android:name=".AndroidPhoneDialer"
android:label="@string/app_name">
            <intent-filter>
                <action android:name="android.intent.action.MAIN" />
```

```
                <category android:name="android.intent.category.LAUNCHER" />
            </intent-filter>
            </activity>
    </application>
<uses-permission android:name="android.permission.CALL_PHONE">
</uses-permission></manifest>
```

The line of most interest is at the end of the file:

```
<uses-permission android:name="android.permission.CALL_PHONE">
</uses-permission>
```

This line of code was added by the Android plugin for Eclipse. If you wanted to, you could have edited AndroidManifest.xml directly to assign the permission. However, if there are times when you are not sure what permission you need to add, or the syntax with which to add it, you can use the Manifest's wizard.

Now that the permission is in place, recompile and run your Activity. Your Emulator should now be making a phone call, as shown in the following illustration.

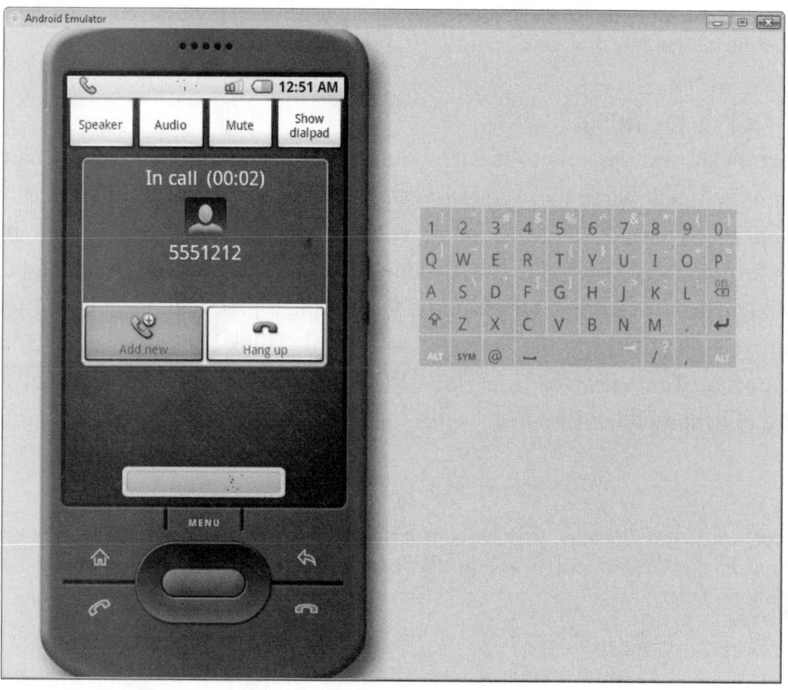

The Activity you created has used an Intent to launch the device's Call Activity and call the number 555-1212. This demonstrates how you can use Intents to your benefit. However, this application does little for you practically. That is to say, how practical would it be to launch an Activity with a hard-coded phone number, just to make a call?

In the following section, you are going to make your application more practical, by adding both a button, to initiate the launching of the Call_Action Intent, and a textbox, to allow the user to input a phone number of their choosing.

Modifying the AndroidPhoneDialer

This section shows you how to modify your AndroidPhoneDialer to make it a bit more practical by adding a few features. By the end of this section, you should be pretty comfortable using not only Intents, but EditTexts and Buttons.

CAUTION

If you did not follow along with the project from the last section, go back and create that Activity. The tutorial in this section assumes that you already have the completed code from the last project at your disposal.

Adding a Button

This section shows you how to modify your project to include a Button. Instead of launching the Intent when the Activity is started, it will be launched by the Button. With the exception of text, buttons are the most prevalent type of object on an application. Buttons form the main interaction between users and applications. Learning how to create and utilize buttons in Android is essential to creating a practical, user-friendly Activity.

You are going to create the Button in main.xml. Think back to Chapter 5, in which you created the TextView for your Hello World! Activity. The TextView had a distinct structure to it, something like this:

NOTE

Remember, when you create a View in main.xml, you are only telling Android what you want the View to look like. You still need to assign functionality to it in AndroidPhoneDialer.java.

```
<View android:id=<id>
android:layout_width=<width>
android:layout_height=<height>
>
```

This formatting is common across views, and the Button is no exception. The XML attributes you need to set for your Button are android:id, android:layout_width, android:layout_height, and android:text. These four XML attributes sufficiently describe your Button so that you can use it within your Activity.

1. Assign to your Button the ID callButton:

```
android:id="@+id/callButton"
```

2. Set layout_width and layout_height to fill_parent and wrap_content, respectively:

```
android:layout_width="fill_parent"
android:layout_height="wrap_content"
```

3. Set the text of the Button to "Show Dialer," which is descriptive enough to identify what the Button will do:

```
android:text="Show Dialer"
```

The full XML for the Button, with attributes, looks like this:

```
<Button android:id="@+id/callButton"
android:layout_width="fill_parent"
android:layout_height="wrap_content"
android:text="Show Dialer" />
```

Take a look now at the finished main.xml file. The Button appears in context and is waiting for you to begin coding it.

```
<?xml version="1.0" encoding="utf-8"?>
<LinearLayout xmlns:android=http://schemas.android.com/apk/res/android
    android:orientation="vertical"
    android:layout_width="fill_parent"
    android:layout_height="fill_parent"
    >
<Button android:id="@+id/callButton"
android:layout_width="fill_parent"
android:layout_height="wrap_content"
android:text="Show Dialer" />
</LinearLayout>
```

To start adding functionality to the Button, you need to add another package to AndroidPhoneDialer.java. The package that contains the Button View is

```
android.widget.Button;
```

You have now imported the Button widget into your project. This gives you the necessary information to begin coding your project. The ultimate outcome of what you are coding in this project should be a Button in your Activity that, when clicked, launches the Call Activity. The Call Activity should be launched with the data "tel:5551212". The resulting screen will match that from your original AndroidPhoneDialer.

The described functionality encompasses a few different concepts. First, you must program a Button that uses the Button attributes you established in main.xml. Next, you have to create a function that will launch the CALL_ACTION Intent code from your previous project. Finally, your Button needs to be able to execute the function and launch the Intent.

The syntax for creating your Button is

```
final Button <button_name> = <button>
```

The left side of this equation creates the Button in your code. The right side of the equation is where you call the Button attributes from main.xml. To call the attributes, you use findViewById( ) and cast the result as a Button. This sounds a little more complicated than it actually is.

Remember, when you added the Button attributes to main.xml, you gave the Button a specific android:id, callButton, which was registered by the Android plugin for Eclipse in the id file as R.id.*callButton*. Use findViewById( ) to retrieve the Button attributes from main.xml by passing it the id callButton:

```
findViewById(R.id.callButton)
```

Don't forget to cast the result as a Button:

```
(Button) findViewById(R.id.callButton)
```

This statement makes up the right side of your equation. The full equation to create your Button looks like this:

```
final Button callButton = (Button) findViewById(R.id.callButton);
```

You now have a Button that you can work with, but you need something for it to do. The Button by itself really does not do much without more code. For purposes of this example, you need to have it execute the CALL_ACTION Intent. Therefore, you are going to create a small function around your existing Intent call. This will give you something to call from when the Button is pressed.

There should be no surprises here if you are familiar with Java programming. You will set up the onClick( ) method to call the Intent code from the previous section. The onClick( ) method takes a View as an argument; however, in this project, you make no calls to the View within the onClick( ) method itself:

```
public void onClick(View v){
      Intent callIntent = new
Intent(Intent.CALL_ACTION,Uri.parse("tel:5551212"));
            callIntent.setLaunchFlags(Intent.NEW_TASK_LAUNCH );
            startActivity(callIntent);

        }
```

The only piece of the application that is left to code is the listener that will tie the Button to the onClick. Listeners should be familiar to Java programmers. For those of you who are not familiar with Java or listeners, *listeners* are the method by which Java objects can “listen” to calls from other objects. The same concept applies within Android. You can establish listeners within Android to let Android Views handle calls from other inputs.

For this project, you need to create for your Button a listener that listens for the onClick event from the Button on the Activity. When the user presses the Button, the listener will call the code in the onClick( ) method. To establish the listener you need, you use the setOnClickListener( ) method of the Button.

If you are familiar with Java development, this structure should not look foreign. This is a typical onClickListener interface implementation in Java. What you will see here is the use of a Java anonymous class to implement the onClickListener for your Button. Also, as an anonymous class, you can make use of local variables—in this case the Button—if that variable is defined as *final*.

The setOnClickListener( ) method takes a pair of arguments. The first is an instantiation of the onClickListener( ). The second is the onClick established earlier. Your setOnClickListener( ) should look like this:

```
callButton.setOnClickListener(new Button.OnClickListener() {
        public void onClick(View v){
              Intent callIntent = new
Intent(Intent.CALL_ACTION,Uri.parse("tel:5551212"));
```

```
            callIntent.setLaunchFlags(Intent.NEW_TASK_LAUNCH );
            startActivity(callIntent);

        }
    });
```

This code segment states that when the callButton is pressed, the onClickListener will execute the code in the onClick. The code in the onClick will execute the CALL_ACTION Intent and call the phone number 555-1212.

Your finished AndroidPhoneDialer.java looks like this:

```
package android_programmers_guide.AndroidPhoneDialer;

import android.app.Activity;
import android.os.Bundle;
import android.widget.Button;
import android.view.View;
import android.content.Intent;
import android.net.Uri;

public class AndroidPhoneDialer extends Activity {
    /** Called when the activity is first created. */
    @Override
    public void onCreate(Bundle icicle) {
        super.onCreate(icicle);
        setContentView(R.layout.main );
      /** Create the Button           */
      final Button callButton = (Button) findViewById(R.id.callButton);
      /** Set the onClickListener to call the onClick  */
      callButton.setOnClickListener(new Button.OnClickListener() {
      /** Use the onClick to call the existing Intent code */
        public void onClick(View v){
              Intent callIntent = new
Intent(Intent.CALL_ACTION,Uri.parse("tel:5551212"));
            callIntent.setLaunchFlags(Intent.NEW_TASK_LAUNCH );
            startActivity(callIntent);
        }
      });
   }
}
```

Compile and run this Activity in the Emulator. The main Activity will display a Button labeled Show Dialer. Click the Button. It should open the Call Activity and dial 555-1212. The main Activity should look as it does in the following illustration.

As you can see, Android is a very robust and flexible platform. With a relatively few lines of code, less than a page, you have created an Activity that utilizes the device's phone hardware and is launched with a Button. Also, by this point you should be pretty comfortable with the way Android handles Activities, Intents, and Views.

Your AndroidPhoneDialer Activity is still rather impractical. You need to add one more item to it. The final section of this chapter shows you how to use the EditText View to let the user input a phone number. The number will then be passed to the CALL_ACTION Intent (instead of the hard-coded value *tel:5551212*).

Implementing an EditText View

You need to add a View to your Activity that will let the user input some text. You will then parse that text and send it to the Intent call from the previous section. Because all Views inherit from the base View, they are helpfully similar in structure and usage. You will find that implementing an EditText is a very simple operation.

First, lay out the Views in your main.xml. You will actually add two Views here: a TextView to act as a label and give some direction to the user, and an EditText to accept the user's input. Together these two Views will add the needed depth and practicality to your Activity.

As you form the look of your Activity, keep in mind that the .xml file is formed visually. This means that if you want the TextView to appear above the EditText on the finished Activity, you should place it before the EditText in main.xml.

Because you have used TextViews a few times now, creation of this View will not get too involved. Simply take a look at the attributes that you set in your TextView:

```
<TextView android:id="@+id/textLabel"
android:layout_width="fill_parent"
android:layout_height="wrap_content"
android:text="Enter Number to Dial:"
/>
```

There is nothing out of the ordinary here. This is just a simple TextView with the text Enter Number to Dial:. This TextView will serve as a label for your EditView. Here's how you set the attributes for the EditView.

```
<EditText android:id="@+id/phoneNumber"
android:layout_width="fill_parent"
android:layout_height="wrap_content"
/>
```

NOTE

You do not have to set the android:text attribute because you do not need any default text.

The id is set to phoneNumber, which is the name you will use to refer to the EditText View in the code. Again, there should be no surprises when setting up main.xml. Your final file should look like this:

```
<?xml version="1.0" encoding="utf-8"?>
<LinearLayout xmlns:android=http://schemas.android.com/apk/res/android
    android:orientation="vertical"
    android:layout_width="fill_parent"
```

```
    android:layout_height="fill_parent"
    >
<TextView android:id="@+id/textLabel"
android:layout_width="fill_parent"
android:layout_height="wrap_content"
android:text="Enter Number to Dial:"
/>
<EditText android:id="@+id/phoneNumber"
android:layout_width="fill_parent"
android:layout_height="wrap_content"
/>
<Button android:id="@+id/callButton"
android:layout_width="fill_parent"
android:layout_height="wrap_content"
android:layout_alignParentRight="true"
android:text="Show Dialer" />

</LinearLayout>
```

The main.xml file is now completed. You can move on to AndroidPhoneDialer.java. If you are not using an existing version of AndroidPhoneDialer.java—one from a previous project in this chapter—you may want to refer to the previous sections to see what code is added to the .java file. This will ensure that you start from the correct point in the code.

The first item you need to add to your .java file is the package definition. You need to add packages not only for the Uri, Button, and Intent, but also for the EditText:

```
import android.widget.Button;
import android.content.Intent;
import android.net.Uri;
import android.widget.EditText;
```

The syntax to set up your EditText View is the same as that for the Button:

```
final EditText <edittext_name> = <edittext>
```

Again, call your EditText phoneNumber. The code to create your EditText is as follows:

```
final EditText phoneNumber = (EditText) findViewById(R.id.phoneNumber);
```

Once your phoneNumber EditText is created, you can use it to reference the text that is input on the device. All you have to do now is call phoneNumber.getText( ) to retrieve the user's input. Replace the hard-coded value "tel:5551212" in the following line,

```
Intent(Intent.CALL_ACTION,Uri.parse("tel:5551212"));
```

with the value of getText( ):

```
Intent(Intent.CALL_ACTION,Uri.parse("tel:" + phoneNumber.getText()));
```

That is all the new code you need to update your project. With these simple two additions, you can give the user an object with which to input a phone number, and send that number to the phone's Call Activity. The full code in the .java file should look like this:

```
package android_programmers_guide.AndroidPhoneDialer;

import android.app.Activity;
import android.os.Bundle;
import android.widget.Button;
import android.view.View;
import android.content.Intent;
import android.net.Uri;
import android.widget.EditText;

public class AndroidPhoneDialer extends Activity {
    /** Called when the activity is first created. */
    @Override
    public void onCreate(Bundle icicle) {
        super.onCreate(icicle);
        setContentView(R.layout.main );
      final EditText phoneNumber = (EditText) findViewById(R.id.phoneNumber
);
      final Button callButton = (Button) findViewById(R.id.callButton);
      callButton.setOnClickListener(new Button.OnClickListener() {
        public void onClick(View v){
              Intent CallIntent = new
Intent(Intent.CALL_ACTION,Uri.parse("tel:" + phoneNumber.getText()));
            CallIntent.setLaunchFlags(Intent.NEW_TASK_LAUNCH );
            startActivity(CallIntent);

        }

      });
    }
}
```

When you run the application in your Emulator, you should see a screen that resembles the following illustration.

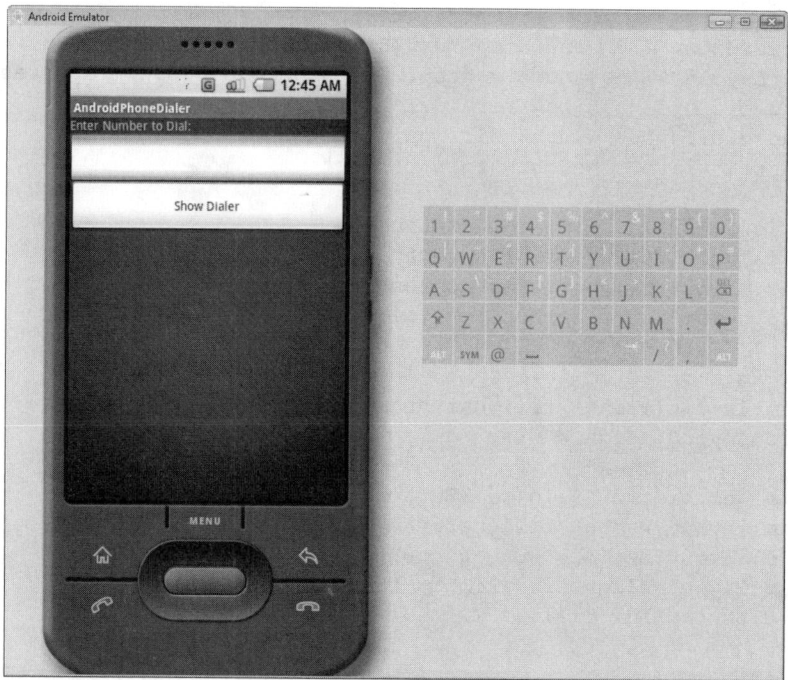

Try This Modify the AndroidPhoneDialer Project

If you played around with the latest version of the AndroidPhoneDialer, you may have noticed something missing. Unfortunately, the way the project is currently written, it allows you to input any type of value into the EditText View and try to send it to the Call Activity. This is really not an optimal approach to application development.

Do some research and add some validation to the EditText. Use the following parameters to modify your project:

- Use a regular expression to validate that a phone number was entered in the EditText (package java.regex).
- Use the showAlert( ) syntax to display a message telling the user they input something that does not match your regular expression.

When you feel you have a working solution, compare it against the following code.

main.xml

```
<?xml version="1.0" encoding="utf-8"?>
<LinearLayout xmlns:android=http://schemas.android.com/apk/res/android
    android:orientation="vertical"
    android:layout_width="fill_parent"
    android:layout_height="fill_parent"
    >
<TextView android:id="@+id/textLabel"
android:layout_width="fill_parent"
android:layout_height="wrap_content"
android:text="Enter Number to Dial:"
/>
<EditText android:id="@+id/phoneNumber"
android:layout_width="fill_parent"
android:layout_height="wrap_content"
/>
<Button android:id="@+id/callButton"
android:layout_width="fill_parent"
android:layout_height="wrap_content"
android:layout_alignParentRight="true"
android:text="Show Dialer" />

</LinearLayout>
```

AndroidPhoneDialer.java

```
package android_programmers_guide.AndroidPhoneDialer;
import android.app.Activity;
import android.os.Bundle;
import android.widget.Button;
import android.view.View;
import android.content.Intent;
import android.net.Uri;
import android.widget.EditText;
import java.util.regex.*;
public class AndroidPhoneDialer extends Activity {
    /** Called when the activity is first created. */
    @Override
    public void onCreate(Bundle icicle) {
        super.onCreate(icicle);
        setContentView(R.layout.main );
        final EditText phoneNumber = (EditText)
findViewById(R.id.phoneNumber );
        final Button callButton = (Button) findViewById(R.id.callButton);
      callButton.setOnClickListener(new Button.OnClickListener() {
```

```
        public void onClick(View v){
              if (validatePhoneNumber(phoneNumber.getText().toString())){
                    Intent CallIntent = new
Intent(Intent.CALL_ACTION,Uri.parse("tel:" + phoneNumber.getText()));
                  CallIntent.setLaunchFlags(Intent.NEW_TASK_LAUNCH );
                  startActivity(CallIntent);
              }
              else{
                    showAlert("Please enter a phone number in the X-XXX-XXX-XXXX
format.",0, "Format Error", "Re-enter Number",false);
              }
        }
      });
   }
        public boolean validatePhoneNumber(String number){
              Pattern phoneNumber = Pattern.compile("(\\d-)?(\\d{3}-)?\\d{3}
\\d{4}");
              Matcher matcher = phoneNumber.matcher(number);
              return matcher.matches();
}
}
```

When you run the project, it should produce a message similar to that in the following illustration.

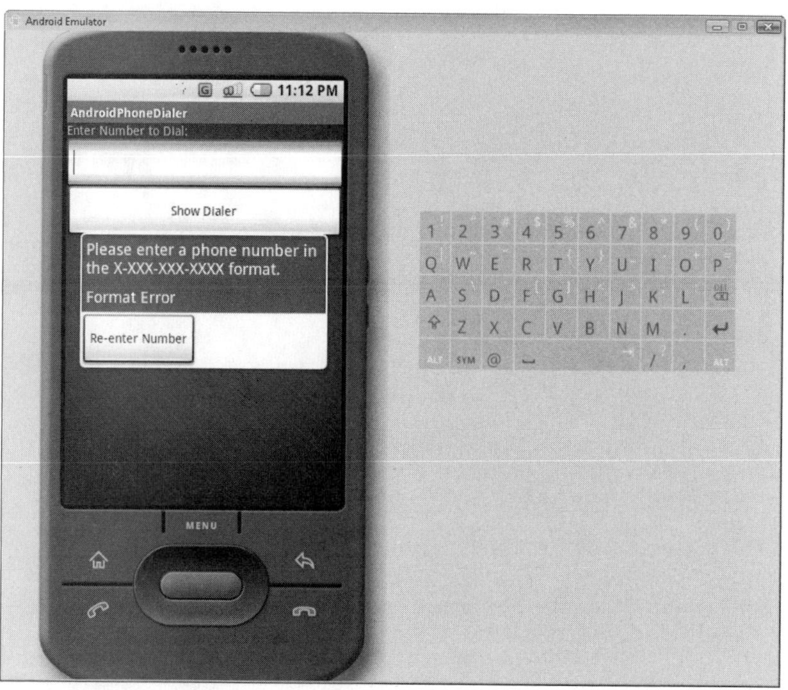

In the next chapter, you will learn about more Views. You will create a multi-Activity application that will allow you to explore and create Views that have not yet been discussed in this book. You will also create and utilize a menu system that will launch your Activities.

Ask the Expert

Q: Is there a way to establish a call to or from the Emulator, to ensure that these Activities are working?

A: As of the time this book was written, there was no way to complete a call to or from the Emulator. However, there was talk from Google that in a future release of the SDK, developers would be able to open two Emulators and complete calls between the two.

Q: Are there other types of Buttons available to Activities—with a different look or feel?

A: Yes. You can use the style attribute to create small Buttons, or small Buttons that include a pointer up, down, left, or right.

Chapter 8

Lists, Menus, and Other Views

Key Skills & Concepts

- Building Activities
- Using Android Menus
- Using the AutoCompleteTextView

This chapter provides a more in-depth treatment of Views and Intents, which arguably are the most important features you should master as a newcomer to Android. These two entities will make up the majority of your early Activities. Almost every Activity you create will have more than one View, and most of them will also need to call an Intent or two.

The best way to learn the most about these topics is to take a very hands-on approach. Reading about these topics and reviewing a list of attributes is one thing, but implementing the code on your own is something entirely different. That is, just as you have done in the previous chapters, you are going to build an Activity that uses Views and Intents rather heavily. By building this application, you will get the best experience with Views and Intents.

The previous two chapters briefly introduced both Views and Intents by having you create very simple Activities that exploited the basic functions of a handful of example Views and Intents. In this chapter, you are going to build a slightly more complex Activity that uses Intents to call new Activities, which you will also create. These new Activities will showcase most of the Views that are available in the current version of the Android SDK. This chapter explains the basic functionality of these Views, such as AutoComplete lists and Galleries, and introduces variations of each View attribute for each Activity.

To begin, create a new Eclipse project named AndroidViews. Create the project with the parameters shown in the following illustration: a Package Name of android_programmers_guide.AndroidViews, an Activity Name of AndroidViews, and an Application Name of AndroidViews.

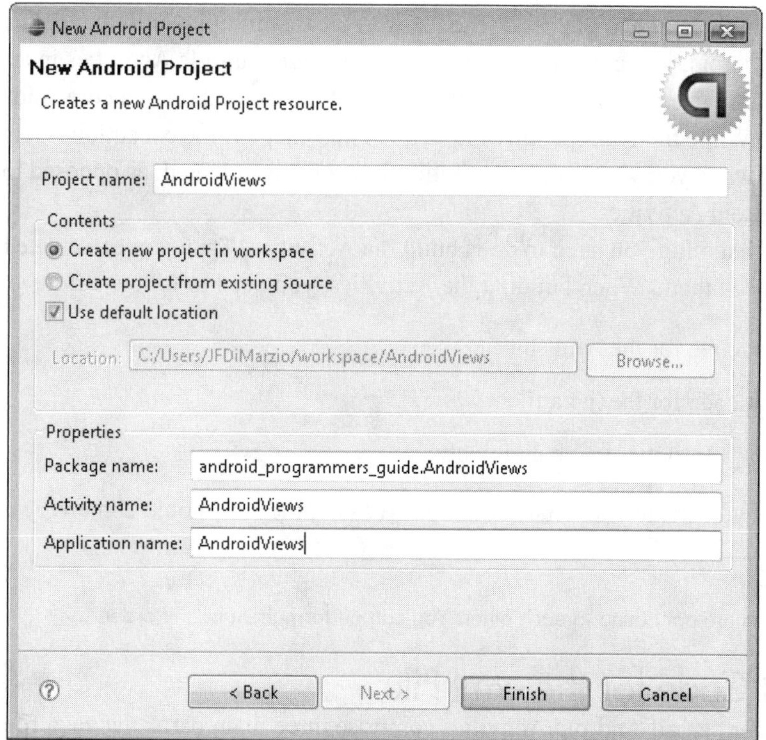

Finish creating the project and open the main.xml file. Remove the Hello World! code from main.xml. With the project created and main.xml cleaned, you can begin to add your code.

Building the Activities

Up to now, you have created only single-Activity applications. This is to say, you have created rather simplistic applications that encompass only one "screen" of data. Take a minute, and think of the last few applications you have used. Chances are, they used more than one "window." Most applications use multiple windows to gather, display, and save data. Your Android applications should be no different.

Although you have not yet learned how to create multiple-Activity applications that run on Android, you got a hint about how to leverage multiple Activities in the last chapter. You used a new concept called Intents to call—and run—a core Android Activity. While the concept still holds true in this chapter, the execution is slightly different when you want to call Activities that you have created, as opposed to calling core Android Activities.

The first thing you need to do is build the Activities. Then you can create the Intents that will call them. When building the Activities, you need to follow a three-step process.

- Intent code for the .xml file
- Intent code for the .java file
- Calling Activities using an Intent

Once you create your first additional Activity, the rest should come very easily.

NOTE

These steps are not bound to each other. You can perform them in any order.

Intent Code for the .xml File

Remember that all Android Activities comprise three main parts: the .java file that contains the code, the .xml file that holds the layout, and the package's Manifest. To this point in the book, you have only used main.xml to control the layout of a single Activity. However, to take advantage of having multiple Activities, you must have multiple .xml layout files.

To create a new .xml file, open your Eclipse project and navigate to the Package Explorer. Open the res directory, right-click the layout folder, and choose New | File.

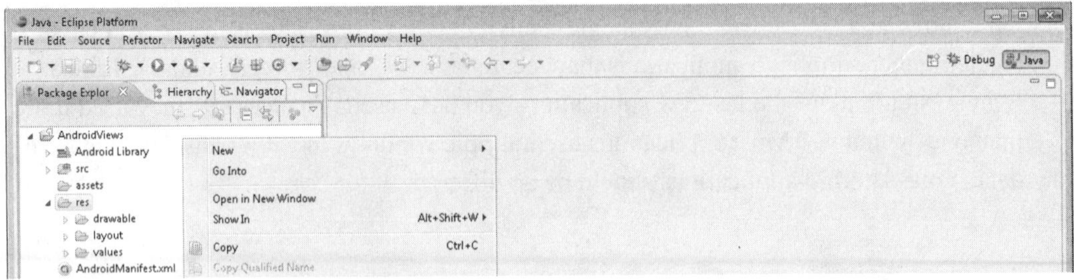

In the New File dialog box, shown next, name your file **test.xml**.

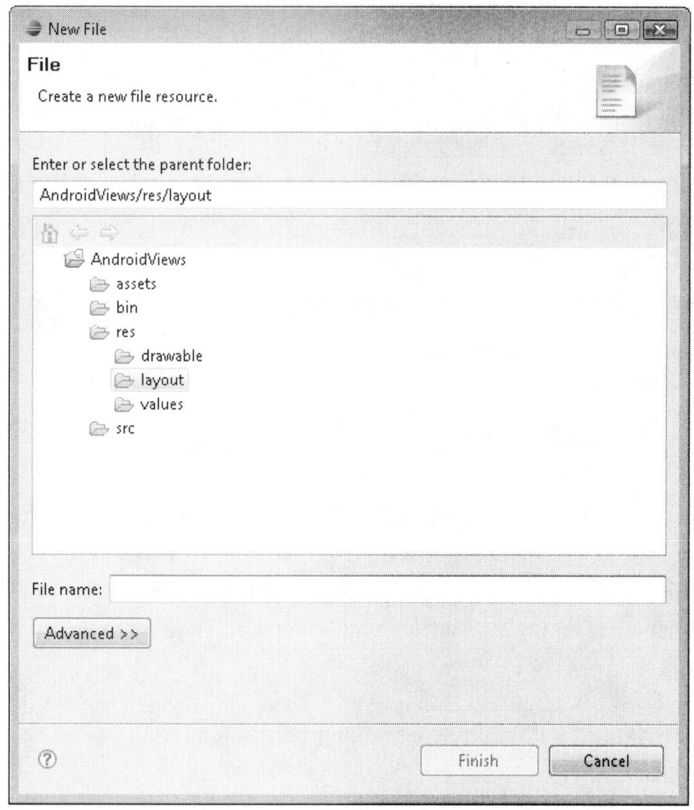

CAUTION

Be sure to enter the filename test.xml in all lowercase letters. New .xml filenames must be all lowercase.

The layout file is created, but it is empty. To get the Activity off on the right foot, add the following code to test.xml. This code will provide a base for your layout. If you need to, you can simply copy this code from the existing main.xml file.

```
<?xml version="1.0" encoding="utf-8"?>
<LinearLayout xmlns:android="http://schemas.android.com/apk/res/android"
    android:orientation="vertical"
    android:layout_width="fill_parent"
    android:layout_height="fill_parent"
    >
</LinearLayout>
```

Intent Code for the .java File

Using the Package Explorer again, navigate to the src directory, open it, and right-click the android_programmers_guide.AndroidViews package, as shown in the following illustration.

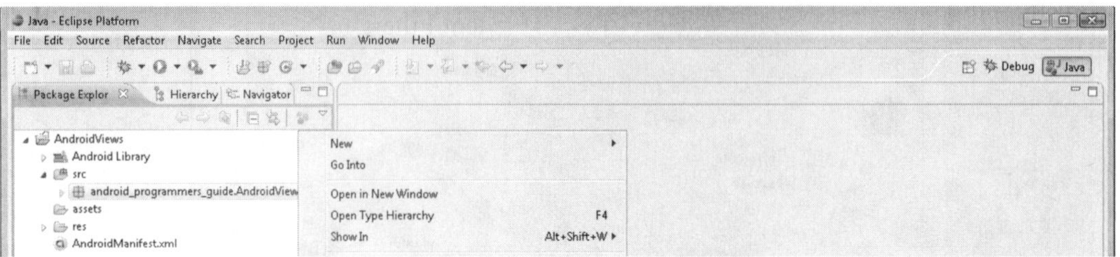

Once again, you are going to add a new file to the folder. After you right-click the AndroidViews package, select New | File from the context menu. This file will hold all the code for the second Activity in this project. Name the file **test.java**. You should now have a nice, new (but empty) .java file. You just need to add a few lines of code to the file to make it usable:

```
package testPackage.test;
import android.app.Activity;
import android.os.Bundle;
public class test extends Activity {
    /** Called when the Activity is first created. */
    @Override
    public void onCreate(Bundle icicle) {
        super.onCreate(icicle);
        setContentView(R.layout.test);
        /** This is our Test Activity
              All code goes below     */
    }
}
```

Notice that you call test.xml in the setContentView method, using the context R.layout.*test*. This line tells the new Activity to use the .xml file you created as the layout file for this "page."

Modifying the AndroidManifest.xml

Open your AndroidManifest.xml file in Eclipse. AndroidManifest.xml has not been discussed in great detail in this book. AndroidManifest.xml contains the global settings for your project. More importantly, AndroidManifest.xml also contains the Intent Filters for your project.

Chapter 7 discussed how Android uses the Intent Filters to marshal what Intents can be accepted by what Activities. The information that facilitates this process is kept in AndroidManifest.xml.

NOTE

There is only one AndroidManifest.xml file per project.

If your AndroidManifest.xml file is currently open, it should appear as follows:

```
<activity android:name=".AndroidViews" android:label="@string/app_name">
  <intent-filter>
    <action android:name="android.intent.action.MAIN" />
    <category android:name="android.intent.category.LAUNCHER" />
   </intent-filter>
</activity>
```

What you are looking at here is the Intent Filter for the AndroidViews Activity, the main Activity that was created with the project. To this file you can add any other Intent Filters that you want your project to handle. In this case, you want to add an Intent Filter that will handle the new Test Activity that you created.

The following is the code for the Intent Filter that you need to add to the AndroidManifest.xml file:

```
<activity android:name=".Test" android:label="Test Activity">
<intent-filter>
   <action android:name="android.intent.action.MAIN" />
   <category android:name="android.intent.category.LAUNCHER" />
</intent-filter>
</activity>
```

Adding this code to AndroidManifest.xml enables Android to pass Intents for the Test Activity to the correct place. The full AndroidManifest.xml file should look like this:

```
<?xml version="1.0" encoding="utf-8"?>
<manifest xmlns:android=http://schemas.android.com/apk/res/android
```

```
    package="android_programmers_guide.AndroidViews">
    <application android:icon="@drawable/icon">
        <activity android:name=".AndroidViews" android:label="@string/app_name">
            <intent-filter>
                <action android:name="android.intent.action.MAIN" />
                <category android:name="android.intent.category.LAUNCHER" />
            </intent-filter>
        </activity>
        <activity android:name=".AutoComplete" android:label="AutoComplete">
           <intent-filter>
                  <action android:name="android.intent.action.MAIN" />
                  <category android:name="android.intent.category.LAUNCHER"
/>
            </intent-filter>
        </activity>
    </application>
</manifest>
```

Now your Activity can handle Intent calls for the Test Activity. To make your Intent call to the Test Activity, you are going to use a structure very similar to the one you used when calling the phone dialer in Chapter 7. The following line of code will set up your Intent:

NOTE

When you start your application, the Activity that will be opened is the AndroidViews Activity that you created with your project. Therefore, place the following code in AndroidViews.java for the purpose of starting the Test Activity.

```
Intent testActivity = new Intent(this, test.class);
```

This line creates an Intent called testActivity. The parameter test.class tells the call that you want the Intent testActivity to represent the Test Activity you created that is associated with *this* Activity.

CAUTION

Do not forget to import the android.content.Intent package when you are working with Intents.

Finally, use the startActivity( ) method to actually start the Test Activity:

```
startActivity(autocomplete);
```

Your completed AndroidViews.java file should look like this:

```
package android_programmers_guide.AndroidViews;

import android.app.Activity;
import android.os.Bundle;
import android.view.Menu;
import android.content.Intent;


public class AndroidViews extends Activity {
    /** Called when the Activity is first created. /
    @Override
    public void onCreate(Bundle icicle) {
        super.onCreate(icicle);
        setContentView(R.layout.main);
        /**Set up our Intent /
```

Run this application in your Android Emulator. Android should launch the AndroidViews Activity, followed quickly by the Test Activity.

In the following section, you will use these techniques to create an application that launches multiple Activities. Each of these Activities will house one View to which you can apply different options. This will give you a great deal of practice displaying and manipulating Views as well as working with Activities.

NOTE

To work with the remaining samples in the upcoming section, remove the Test Activity that you created in this section. You will proceed with the creation of the AndroidViews project without the Test Activity.

Using the Menu

In this section, you are going to build an application that will allow a user to select from a number of different Views. When the user selects a View, a new Activity will be launched containing the selected View.

The tool you are going to use to offer the selections to the user is the Android Menu. Take a look the following illustration. The Menu is displayed when the user activates the Menu Button.

As you can see, selecting the Menu Button from the Android home screen produces a *Wallpaper settings* option. You are going to create a similar menu for your main Activity that will hold all the options for the Views that the user will be able to select from. Right now, the code of your AndroidViews.java file should look like this:

```
package android_programmers_guide.AndroidViews;

import android.app.Activity;
import android.os.Bundle;

public class AndroidViews extends Activity {
    /** Called when the Activity is first created. */
    @Override
    public void onCreate(Bundle icicle) {
        super.onCreate(icicle);
```

```
        setContentView(R.layout.main);
      }
}
```

As with everything you add to your Activities, you need to import a new package to create your menu. Import the android.view.Menu into your AndroidViews Activity:

```
Import android.view.Menu;
```

To create the Menu, you need to override the onCreateOptionsMenu( ) method of the Activity. The method onCreateOptionsMenu( ) is a Boolean method that is called when the user first selects the Menu Button. You will use this method to build your Menu and add your selection items to it. Add the following code to AndroidViews.java:

```
@Override
  public boolean onCreateOptionsMenu(Menu menu) {
 super.onCreateOptionsMenu(menu);
}
```

You will add the code to create the Menu inside the onCreateOptionsMenu( ) method. The items that you need to add to the Menu are the Views that you are going to create in this project. The following is the list of View names that you will need to add to the Menu:

- AutoComplete
- Button
- CheckBox
- EditText
- RadioGroup
- Spinner

In the preceding code that you created to override the onCreateOptionsMenu( ) method, you passed in a Menu variable called *menu*. This variable represents the actual menu item that is created on the Android interface.

To add your list of items to the Menu, you will use the *menu*.add( ) method. The syntax for this call is as follows:

```
menu.add(<group>,<id>,<title>)
```

The parameter group is used to associate the menu items. You will not be using group in this example. However, the value is very important. The parameter id is used to determine what menu item was selected. Finally, the parameter title is the text that will be displayed in the Menu.

Add the following code to the onCreateOptionsMenu( ) method:

```
menu.add(0, 0, "AutoComplete");
menu.add(0, 1, "Button");
menu.add(0, 2, "CheckBox");
menu.add(0, 3, "EditText");
menu.add(0, 4, "RadioGroup");
menu.add(0, 5, "Spinner");
```

Your full AndroidViews.java file should now look like this:

```
package android_programmers_guide.AndroidViews;

import android.app.Activity;
import android.os.Bundle;
import android.view.Menu;

public class AndroidViews extends Activity {
    /** Called when the Activity is first created. */
    @Override
    public void onCreate(Bundle icicle) {
        super.onCreate(icicle);
        setContentView(R.layout.main);
      }

@Override
public boolean onCreateOptionsMenu(Menu menu) {
 super.onCreateOptionsMenu(menu);

 menu.add(0, 0, "AutoComplete");
 menu.add(0, 1, "Button");
 menu.add(0, 2, "CheckBox");
 menu.add(0, 3, "EditText");
 menu.add(0, 4, "RadioGroup");
 menu.add(0, 5, "Spinner");
 return true;
}
}
```

If you execute this code as it is written, you should see the menu shown in Figure 8-1.

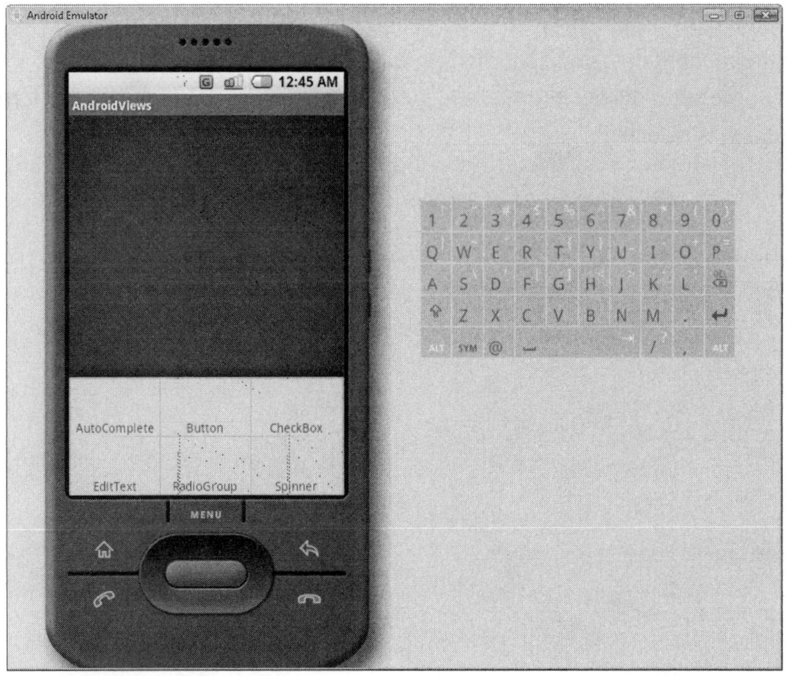

Figure 8-1 Menu with six Views

This is exactly what you wanted to accomplish. However, try clicking any of the options in the Menu. You do not have anything in your Activity that handles events when the user selects a Menu item.

The method you need to override to handle the calls to the Menu items is onOptionsItemSelected( ). Again, like onCreateOptionsMenu( ), onOptionsItemSelected( ) is a Boolean method that you need to override with the specific code to be executed when a Menu item is selected. The override code should look like this:

```
@Override
public boolean onOptionsItemSelected(Menu.Item item){
}
```

There is one problem with this code: onOptionsItemSelected( ) is a general method that is called when *any* menu item is selected. You need to give onOptionsItemSelected( ) a way to differentiate between the menu items and execute code accordingly. Therefore, use a switch/case statement to help the method select between the menu items. When you created the menu items, you specified a series of numbers from 0 to 5 as the values for your menu items. You can use a call to getId( ) in your case statement to determine which menu item was selected:

```
switch (item.getId()) {
 case 0:
     return true;
 case 1:
     return true;
 case 2:
     return true;
 case 3:
     return true;
 case 4:
     return true;
 case 5:
     return true;
 }
 return true;
```

In this case statement, the action for each id is currently set to *return true*. This will not do anything but hold open the area where you need to add code. Your AndroidViews.java file is now ready for use to create Activities that can be launched by the new menu system. The full code of AndroidViews.java should look like this:

```
package android_programmers_guide.AndroidViews;

import android.app.Activity;
import android.os.Bundle;
import android.view.Menu;

public class AndroidViews extends Activity {
    /** Called when the Activity is first created. /
    @Override
    public void onCreate(Bundle icicle) {
        super.onCreate(icicle);
        setContentView(R.layout.main);
       }

@Override
public boolean onCreateOptionsMenu(Menu menu) {
 super.onCreateOptionsMenu(menu);
```

```
 /** Add one menu item for each View in our project */
 menu.add(0, 0, "AutoComplete");
 menu.add(0, 1, "Button");
 menu.add(0, 2, "CheckBox");
 menu.add(0, 3, "EditText");
 menu.add(0, 4, "RadioGroup");
 menu.add(0, 5, "Spinner");
 return true;
}
/** Override onOptionsItemSelected to execute code for each
      menu item */
@Override
public boolean onOptionsItemSelected(Menu.Item item){
}
/** Select statement to handle calls
     to specific menu items */
switch (item.getId()) {
 case 0:
     return true;
 case 1:
     return true;
 case 2:
     return true;
 case 3:
     return true;
 case 4:
     return true;
 case 5:
     return true;
 }
 return true;
}
}
```

With AndroidViews.java complete, you can focus on creating your other Activities. In the following sections, you will create one Activity for each View in your project and add the code to launch that view's Activity in your case statement.

Creating the Activity for AutoComplete

In this section, you are going to create an Activity that will show off the AutoCompleteTextView. AutoCompleteTextViews can be very powerful tools for your applications. This View is specifically helpful at making the most of the limited space available to the Android main screen.

AutoCompleteTextView, as the name implies, is a modified TextView that will refer a possible value for the completion of a word or phrase typed into it. Such Views are greatly useful in mobile applications when you do not want to devote a large amount of space to a ListView, or you want to speed along the process of entering text into your application.

To begin creating your Activity for the AutoCompleteTextView, you need to add a new .xml file for that layout, a .java file for the code, and an Intent Filter to handle the calls.

TIP

The process of creating these items appeared in the "Building the Activities" section earlier in the chapter. Refer to that section as needed to create the following pieces of the project.

Create an autocomplete.xml File

Create a new .xml file in your AndroidViews project named **autocomplete.xml**. Keep in mind that the filename must be all lowercase. The file should appear in the layout folder in your Package Explorer. Double-click the file to edit it.

This file is going to control the layout for your AutoCompleteTextView Activity, so you need to have an AutoCompleteTextView in the layout. The XML for adding an AutoCompleteTextView looks like this:

```
<AutoCompleteTextView android:id="@+id/testAutoComplete"
android:layout_width="fill_parent"
android:layout_height="wrap_content"/>
```

You have created a few Views now in .xml files, so you should be familiar with the format. There is nothing different or unusual about the AutoCompleteTextView. You are setting the id to testAutoComplete, and the width and height to fill_parent and wrap_content, respectively.

You should add the layouts for two Buttons as well. These Buttons will be used to control the attributes that you will change. Name the Buttons **autoCompleteButton** and **textColorButton**, as follows:

```
<Button android:id="@+id/autoCompleteButton"
android:layout_width="fill_parent"
android:layout_height="wrap_content"
android:text="Change Layout"/>

<Button android:id="@+id/textColorButton"
android:layout_width="fill_parent"
```

```
android:layout_height="wrap_content"
android:text="Change Text Color"/>
```

With the three View layouts added, your finished autocomplete.xml file should look like this:

```
<?xml version="1.0" encoding="utf-8"?>
<LinearLayout xmlns:android="http://schemas.android.com/apk/res/android"
    android:orientation="vertical"
    android:layout_width="fill_parent"
    android:layout_height="fill_parent"
    >
<AutoCompleteTextView android:id="@+id/testAutoComplete"
android:layout_width="fill_parent"
android:layout_height="wrap_content"/>
<Button android:id="@+id/autoCompleteButton"
android:layout_width="fill_parent"
android:layout_height="wrap_content"
android:text="Change Layout"/>
<Button android:id="@+id/textColorButton"
android:layout_width="fill_parent"
android:layout_height="wrap_content"
android:text="Change Text Color"/>
</LinearLayout>
```

Create an autocomplete.java File

Follow the instructions that were introduced in the "Creating a New .java File" section earlier in this chapter to create your autocomplete.xml file.

The first thing you need to do is import the packages for your Views. In this Activity, you are using two Views, the AutoCompleteTextView and the Button. You also need to work with Colors and an ArrayAdapter. Therefore, import the following packages with your Activity:

```
package android_programmers_guide.AndroidViews;
import android.app.Activity;
import android.os.Bundle;
import android.view.View;
import android.widget.ArrayAdapter;
import android.widget.AutoCompleteTextView;
import android.widget.Button;
import android.graphics.Color;
```

NOTE

While you may not know what they are for yet, just add the packages for Color and ArrayAdapter. I will explain them later in this section.

Add the initial structure for your AutoComplete class to autocomplete.java:

```
public class AutoComplete extends Activity {
@Override
    public void onCreate(Bundle icicle) {
        super.onCreate(icicle);
}
}
```

This class gives you a base to start building the rest of your Activity. All the functionality of this Activity will be built around this class. The first thing you need to do is load the layout from autocomplete.xml:

```
setContentView(R.layout.autocomplete);
```

For this example, you will create the AutoCompleteTextView so that it contains a list of the months of the year. When a user types into the box, it will anticipate which month they are trying to enter. Given that the AutoCompleteTextView will contain a list of the months, you need to create a list that can be assigned to the AutoCompleteTextView. Create a string array and assign the month values to it:

```
static final String[] Months = new String[]{
"January","February","March","April","May","June","July","August",
"September","October","November","December"
};
```

The next task is to assign this string array to the AutoCompleteTextView. You have created more than a few Views by now, so the code to create the AutoCompleteTextView should look very familiar. What you have not seen before is the code to assign the string array to the View:

```
ArrayAdapter<String> monthArray = new ArrayAdapter<String>(this,
                android.R.layout.simple_list_item_1, Months);
final AutoCompleteTextView textView = (AutoCompleteTextView)
findViewById(R.id.testAutoComplete);
textView.setAdapter(monthArray);
```

In the first line, you are taking the string array you created and assigning it to an ArrayAdapter named monthArray. Next, you instantiate your AutoCompleteTextView by locating it in the .xml layout file. Finally, you use the setAdapter( ) method to assign the monthArray ArrayAdapter to the AutoCompleteTextView.

The next snippet of code instantiates the two Buttons. This is the same code you have used in previous chapters. The only difference here from other code you have written is that you are calling two functions, changeOption and changeOption2, which have not been created yet.

NOTE

Notice that you are passing the AutoCompleteTextView into the function calls. You will need to create this parameter when you create the functions.

```
final Button changeButton = (Button) findViewById(R.id.autoCompleteButton);
        changeButton.setOnClickListener(new Button.OnClickListener() {
              public void onClick(View v){
                    changeOption(textView);
              }
        });
        final Button changeButton2 = (Button)
findViewById(R.id.textColorButton);
        changeButton2.setOnClickListener(new Button.OnClickListener() {
              public void onClick(View v){
                    changeOption2(textView);
              }
        });
```

The functions called by these Buttons will be used to change layout attributes on the AutoCompleteTextView. The two attributes I have chosen to modify (using the Buttons) are the layout height and text color. You are going to set up one of these Buttons to change the AutoCompleteTextView's layout height from 30 to 100 and back. The other Button will change the color of the text within the View to red.

The function changeOption( ) will change the AutoCompleteTextView's layout height. The code for this function is very simple:

```
public void changeOption(AutoCompleteTextView text){
      if (text.getHeight()==100){
      text.setHeight(30);
      }
      else{
           text.setHeight(100);
      }
}
```

What you are doing in this function is checking the current height of the AutoCompleteTextView. If that height is 100, you set it to 30; otherwise you set it to 100.

The changeOption2( ) function is just as easy:

```
public void changeOption2(AutoCompleteTextView text){
text.setTextColor(Color.RED);
}
}
```

This function simply sets the text color of the AutoCompleteTextView to Color.RED. The value Color.RED is imported from the android.graphics.Color package. You can browse this package and change the color to any value; I selected RED so that it would stand out.

Your full autocomplete.java file should now look like this:

```
package android_programmers_guide.AndroidViews;

import android.app.Activity;
import android.os.Bundle;
import android.view.View;
import android.widget.ArrayAdapter;
import android.widget.AutoCompleteTextView;
import android.widget.Button;
import android.graphics.Color;

public class AutoComplete extends Activity {
    @Override
    public void onCreate(Bundle icicle) {
        super.onCreate(icicle);
        setContentView(R.layout.autocomplete);

        ArrayAdapter<String> monthArray = new ArrayAdapter<String>(this,
                android.R.layout.simple_list_item_1, Months);
        final AutoCompleteTextView textView = (AutoCompleteTextView)
findViewById(R.id.testAutoComplete);
        textView.setAdapter(monthArray);
        final Button changeButton = (Button)
findViewById(R.id.autoCompleteButton);
        changeButton.setOnClickListener(new Button.OnClickListener() {
              public void onClick(View v){
                    changeOption(textView);

              }
        });
        final Button changeButton2 = (Button)
findViewById(R.id.textColorButton);
```

```
        changeButton2.setOnClickListener(new Button.OnClickListener() {
              public void onClick(View v){
                    changeOption2(textView);

              }
        });
    }

static final String[] Months = new String[]{
"January","February","March","April","May","June","July","August",
"September","October","November","December"
};

public void changeOption(AutoCompleteTextView text){
      if (text.getHeight()==100){
      text.setHeight(30);
      }
      else{
           text.setHeight(100);

      }
}
public void changeOption2(AutoCompleteTextView text){
text.setTextColor(Color.RED);
}
}
```

Add an Intent Filter

The last thing you need to do before you can run this application is to set up the Intent Filter in AndroidManifest.xml. You will then be able to call that Intent from the Menu shown earlier in Figure 8-1. The code for the Intent Filter should look as follows:

```
<activity android:name=".AutoComplete" android:label="AutoComplete">
      <intent-filter>
      <action android:name="android.intent.action.MAIN" />
      <category android:name="android.intent.category.LAUNCHER" />
      </intent-filter>
</activity>
```

Here is your completed AndroidManifest.xml file for this project:

```
<?xml version="1.0" encoding="utf-8"?>
<manifest xmlns:android=http://schemas.android.com/apk/res/android
    package="android_programmers_guide.AndroidViews">
```

```
    <application android:icon="@drawable/icon">
        <activity android:name=".AndroidViews"
android:label="@string/app_name">
            <intent-filter>
                <action android:name="android.intent.action.MAIN" />
                <category android:name="android.intent.category.LAUNCHER" />
            </intent-filter>
        </activity>
        <activity android:name=".AutoComplete" android:label="AutoComplete">
            <intent-filter>
                  <action android:name="android.intent.action.MAIN" />
                  <category android:name="android.intent.category.LAUNCHER"
/>
            </intent-filter>
        </activity>
    </application>
</manifest>
```

Handling the Intent Call

With AndroidManifest.xml complete, add the following function to AndroidViews.java:

```
public void showAutoComplete(){
      Intent autocomplete = new Intent(this, AutoComplete.class);
      startActivity(autocomplete);
}
```

When called from your select/case statement, this function will open your autocomplete Activity. Edit the case 0 of the select statement to let it call the new function:

```
case 0:
     showAutoComplete();
     return true;
```

Run the application in the Android Emulator. When the main Activity is launched, click the Menu Button, and you should see the menu shown earlier in Figure 8-1. Click the AutoComplete menu item.

Clicking the AutoComplete Button menu item should bring up your autocomplete Activity, shown next.

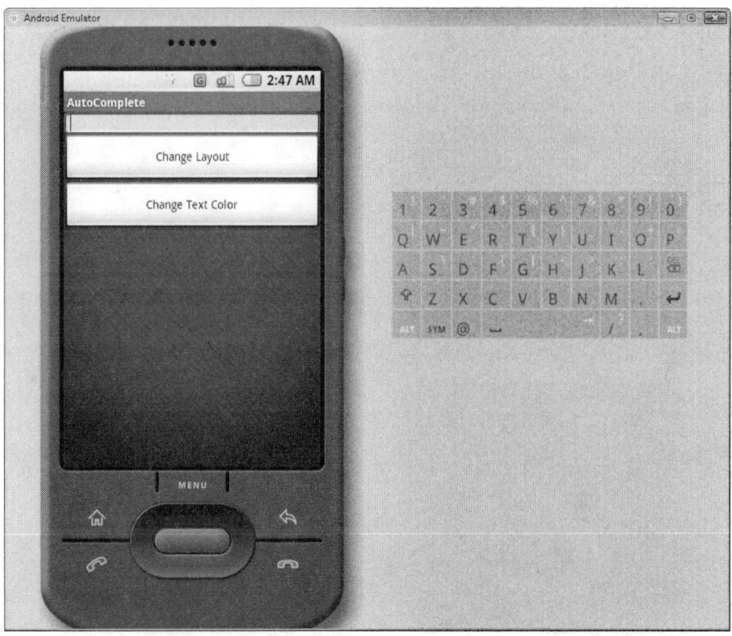

To test the AutoCompleteTextView, begin typing the word **January**. After you type a few characters, you should see the word January appear under the text box, as shown in the following illustration.

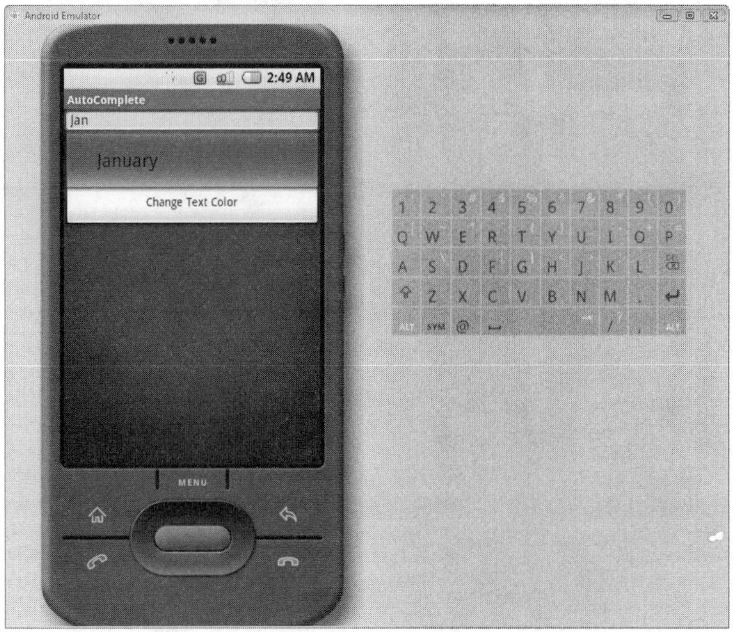

Next, click the Change Layout Button, the result of which should be an expanded text entry box similar to that shown in the following illustration.

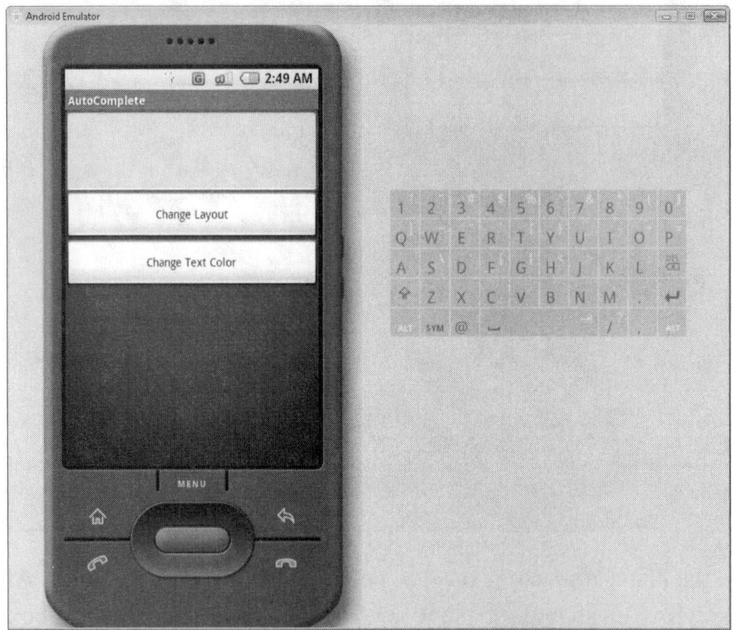

Now click the Change Text Color Button and type some text, as shown next.

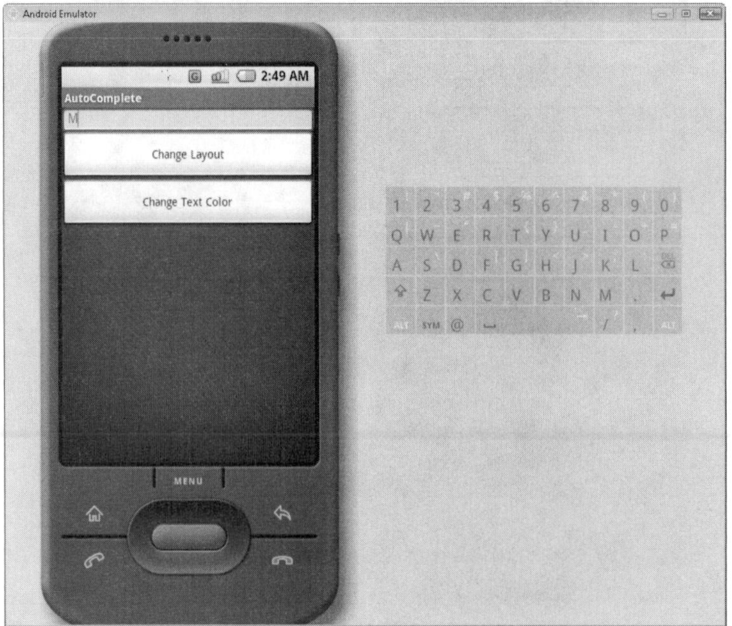

The following sections give you the supporting code for implementing the remaining five Views in your project.

Button

Take a look at the following code. This code represents four files, AndroidManifest.xml, Button.xml, testButton.java, and AndroidViews.java. Add the code in these files to your existing AndroidViews Activity.

CAUTION

If you have not followed this chapter from the beginning, you may have trouble implementing this code. To make sure you are getting the full project, follow this chapter from the beginning.

AndroidManifest.xml

```
<?xml version="1.0" encoding="utf-8"?>
<manifest xmlns:android=http://schemas.android.com/apk/res/android
    package="android_programmers_guide.AndroidViews"
    <application android:icon="@drawable/icon">
        <activity android:name=".AndroidViews"
android:label="@string/app_name">
            <intent-filter>
                <action android:name="android.intent.action.MAIN" />
                <category android:name="android.intent.category.LAUNCHER" />
            </intent-filter>
        </activity>
        <activity android:name=".AutoComplete" android:label="AutoComplete">
            <intent-filter>
                  <action android:name="android.intent.action.MAIN" />
                  <category android:name="android.intent.category.LAUNCHER"/>
            </intent-filter>
        </activity>
         <activity android:name=".testButton" android:label="TestButton">
            <intent-filter>
                  <action android:name="android.intent.action.MAIN" />
                  <category android:name="android.intent.category.LAUNCHER"/>
            </intent-filter>
        </activity>
    </application>
</manifest>
```

Button.xml

```
<?xml version="1.0" encoding="utf-8"?>
<LinearLayout xmlns:android=http://schemas.android.com/apk/res/android
    android:orientation="vertical"
    android:layout_width="fill_parent"
```

```
    android:layout_height="fill_parent">
<Button android:id="@+id/testButton"
android:layout_width="fill_parent"
android:layout_height="wrap_content"
android:text="This is the test Button"/>
<Button android:id="@+id/layoutButton"
android:layout_width="fill_parent"
android:layout_height="wrap_content"
android:text="Change Layout"/>
<Button android:id="@+id/textColorButton"
android:layout_width="fill_parent"
android:layout_height="wrap_content"
android:text="Change Text Color"/>
</LinearLayout>
```

testButton.java

```
package android_programmers_guide.AndroidViews;

import android.app.Activity;
import android.os.Bundle;
import android.view.View;
import android.widget.Button;
import android.graphics.Color;
public class testButton extends Activity {
    @Override
    public void onCreate(Bundle icicle) {
        super.onCreate(icicle);
        setContentView(R.layout.Button);

        final Button Button = (Button) findViewById(R.id.testButton);

        final Button changeButton = (Button)findViewById(R.id.layoutButton);
        changeButton.setOnClickListener(new Button.OnClickListener() {
            public void onClick(View v){
                changeOption(Button); }
        });
        final Button changeButton2 = (Button)
findViewById(R.id.textColorButton);
        changeButton2.setOnClickListener(new Button.OnClickListener() {
            public void onClick(View v){
                changeOption2(Button);
            }
        });
    }
    public void changeOption(Button Button){
      if (Button.getHeight()==100){
           Button.setHeight(30);
      }
```

```
      else{
            Button.setHeight(100);
      }
    }
    public void changeOption2(Button Button){
      Button.setTextColor(Color.RED);
    }
}
```

AndroidViews.java

```
package android_programmers_guide.AndroidViews;

import android.app.Activity;
import android.os.Bundle;
import android.view.Menu;
import android.content.Intent;

public class AndroidViews extends Activity {
    /** Called when the Activity is first created. */
    @Override
    public void onCreate(Bundle icicle) {
        super.onCreate(icicle);
        setContentView(R.layout.main);
       }
@Override
public boolean onCreateOptionsMenu(Menu menu) {
 super.onCreateOptionsMenu(menu);

 menu.add(0, 0, "AutoComplete");
 menu.add(0, 1, "Button");
 menu.add(0, 2, "CheckBox");
 menu.add(0, 3, "EditText");
 menu.add(0, 4, "RadioGroup");
 menu.add(0, 5, "Spinner");
 return true;
}
@Override
public boolean onOptionsItemSelected(Menu.Item item){
 switch (item.getId()) {
 case 0:
     showAutoComplete();
     return true;
 case 1:
     showButton();
     return true;
```

```
 case 2:
     return true;
 case 3:
     return true;
 case 4:
     return true;
 case 5:
     return true;
 }
 return true;
}
public void showButton() {
      Intent showButton = new Intent(this, testButton.class);
      startActivity(showButton);
}
public void showAutoComplete(){
      Intent autocomplete = new Intent(this, AutoComplete.class);
      startActivity(autocomplete);
}
}
```

Launch your application and select the Button option from the Menu (shown earlier in Figure 8-1).

The following illustration shows what the Button Activity looks like.

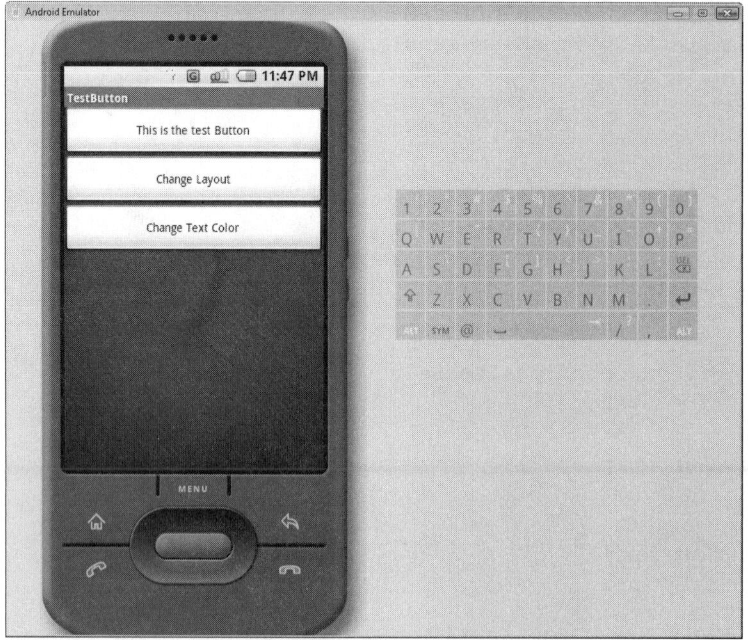

Try clicking the Change Layout Button. Again, the result is a wider display area for the text, as depicted in the following illustration.

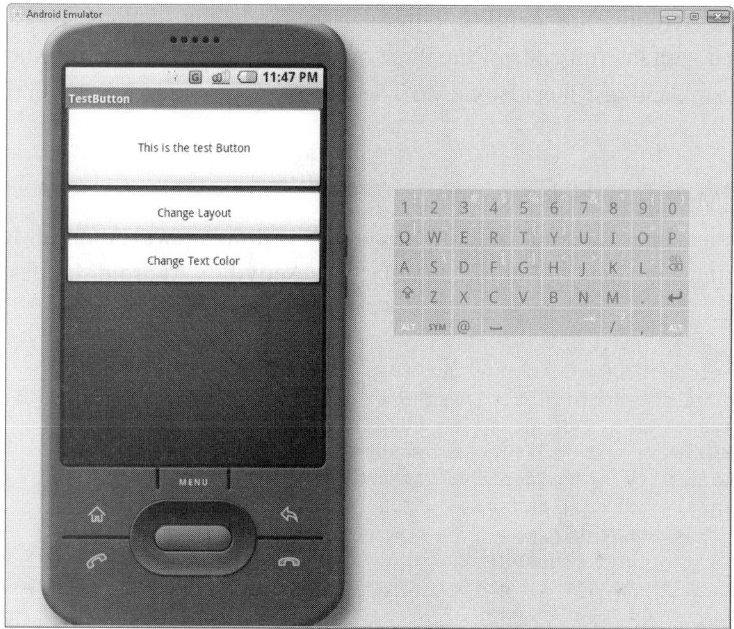

Click the Change Text Color Button and the text should turn red, as shown next.

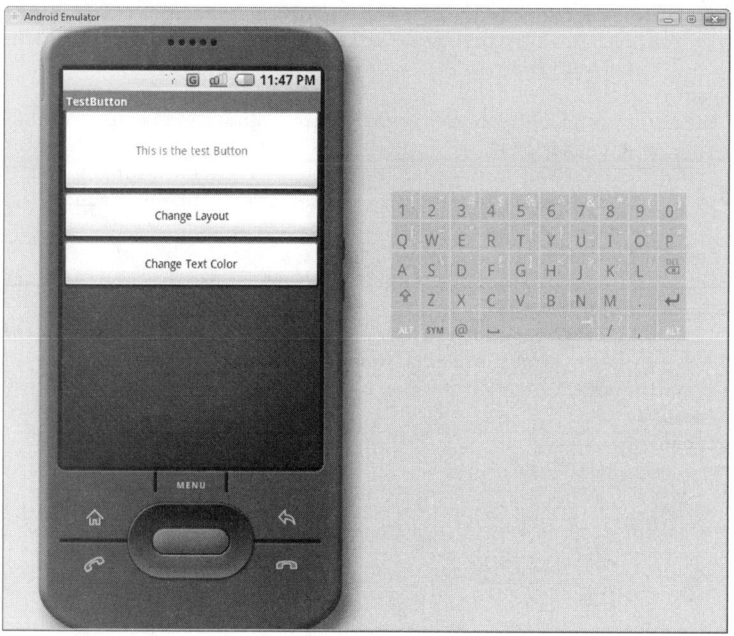

CheckBox

In this section you will be creating an Activity for the CheckBox View. The steps for creating the Activities are identical to those in the preceding sections. Therefore, you will be provided with the full code of the three main Activity files—AndroidManifest.xml, checkbox.xml, and testCheckBox.java. These files are provided for you in the following sections.

AndroidManifest.xml

This section contains the full code of the current AndroidViews' AndroidManifest.xml. If you are following along in Eclipse, modify your Activity's AndroidManifest.xml to look as follows:

```
<?xml version="1.0" encoding="utf-8"?>
<manifest xmlns:android=http://schemas.android.com/apk/res/android
    package="android_programmers_guide.AndroidViews">
    <application android:icon="@drawable/icon">
        <activity android:name=".AndroidViews"
android:label="@string/app_name">
            <intent-filter>
                <action android:name="android.intent.action.MAIN" />
                <category android:name="android.intent.category.LAUNCHER" />
            </intent-filter>
        </activity>
        <activity android:name=".AutoComplete" android:label="AutoComplete">
            <intent-filter>
                  <action android:name="android.intent.action.MAIN" />
            <category android:name="android.intent.category.LAUNCHER" />
            </intent-filter>
        </activity>
         <activity android:name=".testButton" android:label="TestButton">
            <intent-filter>
                  <action android:name="android.intent.action.MAIN" />
                  <category android:name="android.intent.category.LAUNCHER"/>
            </intent-filter>
        </activity>
        <activity android:name=".testCheckBox" android:label="TestCheckBox">
            <intent-filter>
                  <action android:name="android.intent.action.MAIN" />
                  <category android:name="android.intent.category.LAUNCHER"/>
            </intent-filter>
        </activity>
    </application>
</manifest>
```

checkbox.xml

This section shows the complete code of the checkbox.xml. Create a new XML file in your project named checkbox.xml using the instructions outlined earlier in this chapter. Use the following code to model your file.

```
<?xml version="1.0" encoding="utf-8"?>
<LinearLayout xmlns:android=http://schemas.android.com/apk/res/android
    android:orientation="vertical"
    android:layout_width="fill_parent"
    android:layout_height="fill_parent"
    >
<CheckBox android:id="@+id/testCheckBox"
android:layout_width="fill_parent"
android:layout_height="wrap_content"
android:text="This is the test CheckBox"/>
<Button android:id="@+id/layoutButton"
android:layout_width="fill_parent"
android:layout_height="wrap_content"
android:text="Change Layout"/>
<Button android:id="@+id/textColorButton"
android:layout_width="fill_parent"
android:layout_height="wrap_content"
android:text="Change Text Color"/>
</LinearLayout>
```

testCheckBox.java

This section covers the final new file needed to implement your CheckBox Activity. Create a new .java file in your project named testCheckBox.java. This file is the main file of the Activity and contains the actionable code. Use the following code in your testCheckBox.java.

```
package android_programmers_guide.AndroidViews;

import android.app.Activity;
import android.os.Bundle;
import android.view.View;
import android.widget.CheckBox;
import android.widget.Button;
import android.graphics.Color;

public class testCheckBox extends Activity {
```

```
    @Override
    public void onCreate(Bundle icicle) {
        super.onCreate(icicle);
        setContentView(R.layout.checkbox);

        final CheckBox checkbox = (CheckBox)findViewById(R.id.testCheckBox);

        final Button changeButton = (Button)findViewById(R.id.layoutButton);
        changeButton.setOnClickListener(new Button.OnClickListener() {
              public void onClick(View v){
                    changeOption(checkbox); }
        });
        final Button changeButton2 = (Button)
findViewById(R.id.textColorButton);
        changeButton2.setOnClickListener(new Button.OnClickListener() {
              public void onClick(View v){
                    changeOption2(checkbox);
              }
        });
    }
    public void changeOption(CheckBox checkbox){
      if (checkbox.getHeight()==100){
            checkbox.setHeight(30);
      }
      else{
            checkbox.setHeight(100);
            }
    }
    public void changeOption2(CheckBox checkbox){
      checkbox.setTextColor(Color.RED);
    }
}
```

AndroidViews.java

The last step to create this Activity is to edit the AndroidViews.java. If you want to call the testCheckBox Activity from the main AndroidViews Activity, you must add code to the AndroidViews.java. Compare the following code with that in your current AndroidViews.java. Add the needed code to complete your file.

```
package android_programmers_guide.AndroidViews;

import android.app.Activity;
import android.os.Bundle;
import android.view.Menu;
import android.content.Intent;
```

```
public class AndroidViews extends Activity {
    /** Called when the Activity is first created. */
    @Override
    public void onCreate(Bundle icicle) {
        super.onCreate(icicle);
        setContentView(R.layout.main);
       }
@Override
public boolean onCreateOptionsMenu(Menu menu) {
 super.onCreateOptionsMenu(menu);

 menu.add(0, 0, "AutoComplete");
 menu.add(0, 1, "Button");
 menu.add(0, 2, "CheckBox");
 menu.add(0, 3, "EditText");
 menu.add(0, 4, "RadioGroup");
 menu.add(0, 5, "Spinner");
 return true;
}
@Override
public boolean onOptionsItemSelected(Menu.Item item){
 switch (item.getId()) {
 case 0:
     showAutoComplete();
     return true;
 case 1:
     showButton();
     return true;
 case 2:
       showCheckBox()
       return true;
 case 3:
       return true;
 case 4:
       return true;
 case 5:
       return true;
 }
 return true;
}
public void showButton() {
      Intent showButton = new Intent(this, testButton.class);
      startActivity(showButton);
```

```
}
public void showAutoComplete(){
      Intent autocomplete = new Intent(this, AutoComplete.class);
      startActivity(autocomplete);
}
public void showCheckBox() {
      Intent checkbox = new Intent(this, testCheckBox.class);
      startActivity(checkbox);
 }
}
```

Launch your application and select the CheckBox option from the Menu (shown earlier in Figure 8-1).

The following illustration shows what the CheckBox Activity looks like.

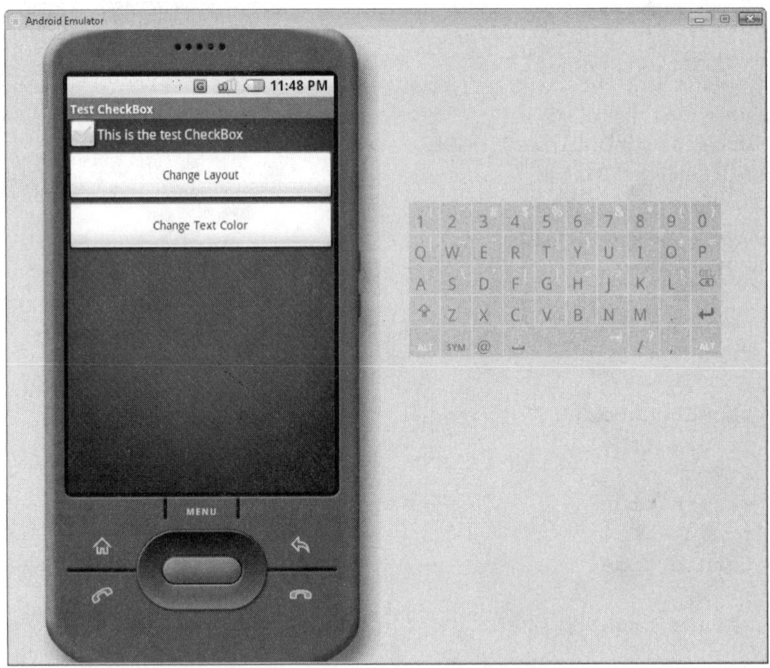

Try clicking the Change Layout and Change Test Color Buttons. The results are depicted in the following illustrations.

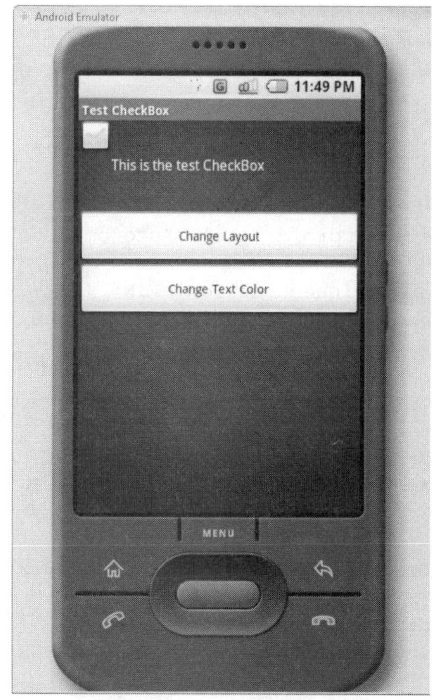

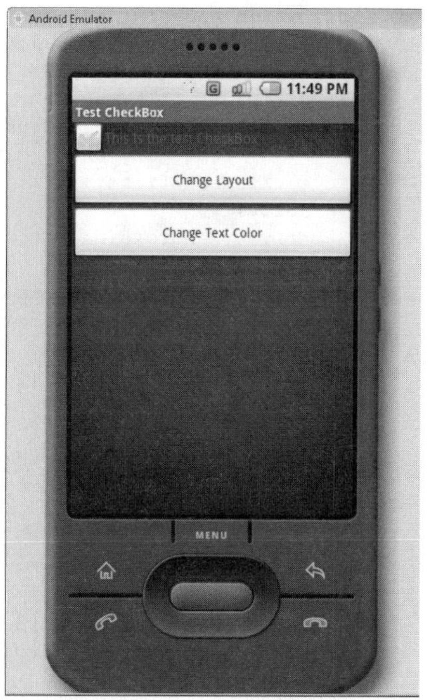

EditText

In this section, much like the last, you will be creating an Activity for the EditText View. The steps for creating the Activities are identical to those in the preceding sections. Therefore you will be provided with the full code of the three main Activity files—AndroidManifest.xml, edittext.xml, and testEditText.java. These files are provided for you in the following sections.

AndroidManifest.xml

This section contains the full code of the current AndroidViews' AndroidManifest.xml. If you are following along in Eclipse, modify your Activity's AndroidManifest.xml to look as follows:

```
<?xml version="1.0" encoding="utf-8"?>
<manifest xmlns:android=http://schemas.android.com/apk/res/android
    package="android_programmers_guide.AndroidViews">
```

```
    <application android:icon="@drawable/icon">
        <activity android:name=".AndroidViews"
android:label="@string/app_name">
            <intent-filter>
                <action android:name="android.intent.action.MAIN" />
                <category android:name="android.intent.category.LAUNCHER" />
            </intent-filter>
        </activity>
        <activity android:name=".AutoComplete" android:label="AutoComplete">
            <intent-filter>
                  <action android:name="android.intent.action.MAIN" />
                  <category android:name="android.intent.category.LAUNCHER"/>
            </intent-filter>
        </activity>
         <activity android:name=".testButton" android:label="TestButton">
            <intent-filter>
                  <action android:name="android.intent.action.MAIN" />
                  <category android:name="android.intent.category.LAUNCHER"/>
            </intent-filter>
        </activity>
        <activity android:name=".testCheckBox" android:label="TestCheckBox">
            <intent-filter>
                  <action android:name="android.intent.action.MAIN" />
                  <category android:name="android.intent.category.LAUNCHER"/>
            </intent-filter>
        </activity>
        <activity android:name=".testEditText" android:label="TestEditText">
            <intent-filter>
                  <action android:name="android.intent.action.MAIN" />
                  <category android:name="android.intent.category.LAUNCHER"/>
            </intent-filter>
        </activity>
    </application>
</manifest>
```

edittext.xml

This section shows the complete code of the edittext.xml. Create a new XML file in your project named edittext.xml using the instructions outlined earlier in this chapter. Use the following code to model your file.

```
<?xml version="1.0" encoding="utf-8"?>
<LinearLayout xmlns:android=http://schemas.android.com/apk/res/android
    android:orientation="vertical"
    android:layout_width="fill_parent"
```

```
    android:layout_height="fill_parent"
    >
<EditText android:id="@+id/testEditText"
android:layout_width="fill_parent"
android:layout_height="wrap_content"
/>
<Button android:id="@+id/layoutButton"
android:layout_width="fill_parent"
android:layout_height="wrap_content"
android:text="Change Layout"/>
<Button android:id="@+id/textColorButton"
android:layout_width="fill_parent"
android:layout_height="wrap_content"
android:text="Change Text Color"/>
</LinearLayout>
```

testEditText.java

This section covers the final file needed to implement your EditText Activity. Create a new .java file in your project named testEditText.java. This file is the main file of the Activity and contains the actionable code. Use the following code in your testEditText.java to finish this Activity.

```
package android_programmers_guide.AndroidViews;

import android.app.Activity;
import android.os.Bundle;
import android.view.View;
import android.widget.EditText;
import android.widget.Button;
import android.graphics.Color;

public class testEditText extends Activity {
    @Override
    public void onCreate(Bundle icicle) {
        super.onCreate(icicle);
        setContentView(R.layout.edittext);

        final EditText edittext = (EditText)findViewById(R.id.testEditText);

        final Button changeButton = (Button)findViewById(R.id.layoutButton);
        changeButton.setOnClickListener(new Button.OnClickListener() {
             public void onClick(View v){
```

```
                    changeOption(edittext); }
        });
        final Button changeButton2 = (Button)
findViewById(R.id.textColorButton);
        changeButton2.setOnClickListener(new Button.OnClickListener() {
              public void onClick(View v){
                    changeOption2(edittext);
              }
        });
    }
    public void changeOption(EditText edittext){
      if (edittext.getHeight()==100){
            edittext.setHeight(30);
      }
      else{
            edittext.setHeight(100);
            }
    }
    public void changeOption2(EditText edittext){
      edittext.setTextColor(Color.RED);
    }
}
```

AndroidViews.java

The last step to create this Activity is to edit the AndroidViews.java. If you want to call the testEditText Activity from the main AndroidViews Activity, you must add code to the AndroidViews.java. Compare the following code with that in your current AndroidViews.java. Add the needed code to complete your file.

```
package android_programmers_guide.AndroidViews;

import android.app.Activity;
import android.os.Bundle;
import android.view.Menu;
import android.content.Intent;

public class AndroidViews extends Activity {
    /** Called when the Activity is first created. */
    @Override
    public void onCreate(Bundle icicle) {
        super.onCreate(icicle);
        setContentView(R.layout.main);
       }
```

```
@Override
public boolean onCreateOptionsMenu(Menu menu) {
 super.onCreateOptionsMenu(menu);

 menu.add(0, 0, "AutoComplete");
 menu.add(0, 1, "Button");
 menu.add(0, 2, "CheckBox");
 menu.add(0, 3, "EditText");
 menu.add(0, 4, "RadioGroup");
 menu.add(0, 5, "Spinner");
 return true;
}
@Override
public boolean onOptionsItemSelected(Menu.Item item){
 switch (item.getId()) {
 case 0:
     showAutoComplete();
     return true;
 case 1:
     showButton();
     return true;
 case 2:
     showCheckBox();
     return true;
 case 3:
     showEditText();
     return true;
 case 4:
     showRadioGroup();
     return true;
 case 5:
     showSpinner();
     return true;
 }
 return true;
}
public void showButton() {
      Intent showButton = new Intent(this, testButton.class);
      startActivity(showButton);
}
public void showAutoComplete(){
      Intent autocomplete = new Intent(this, AutoComplete.class);
```

```
        startActivity(autocomplete);
    }
    public void showCheckBox(){
        Intent checkbox = new Intent(this, testCheckBox.class);
        startActivity(checkbox);
     }
    public void showEditText() {
        Intent edittext = new Intent(this, testEditText.class);
        startActivity(edittext);
     }
    }
```

Launch your application and select the EditText option from the Menu (shown earlier in Figure 8-1).

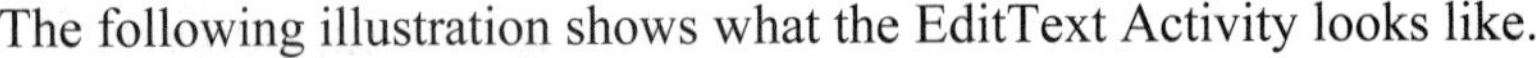

The following illustration shows what the EditText Activity looks like.

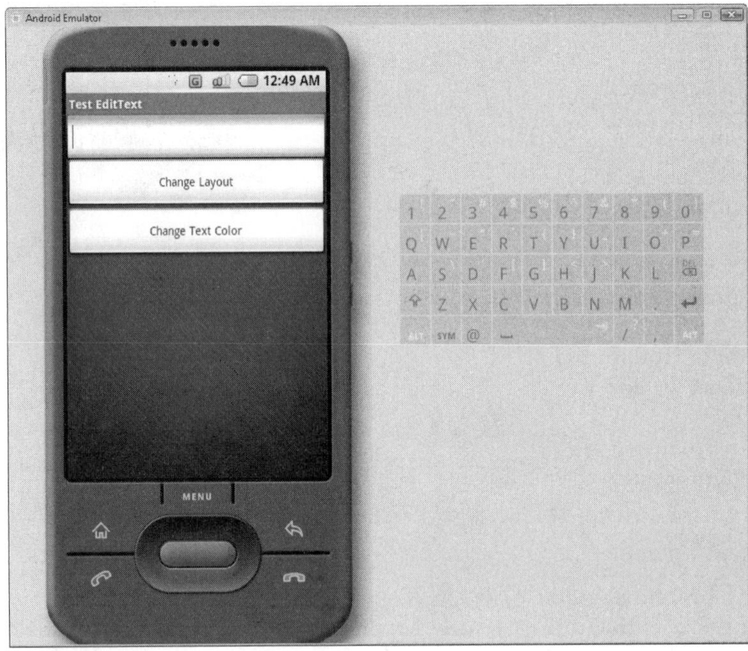

Click the Change Layout and Change Test Color Buttons. The results are depicted in the following illustrations.

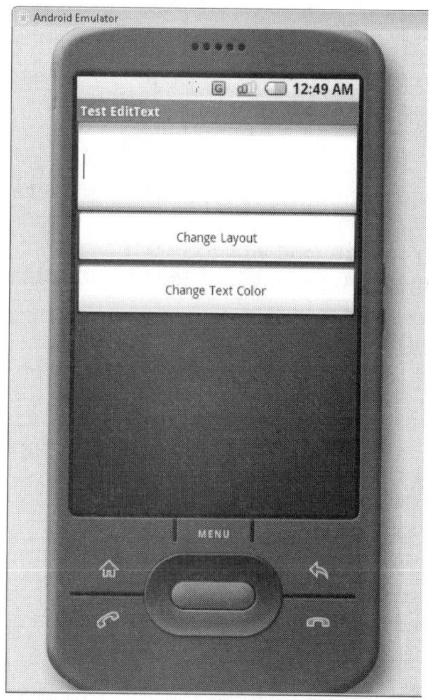

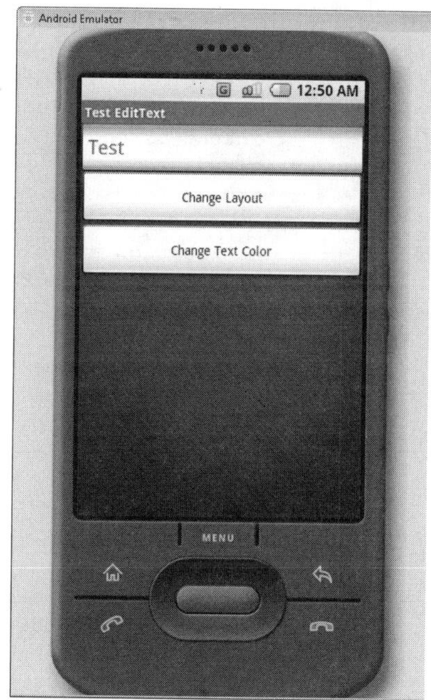

RadioGroup

In this section you will be creating an Activity for the RadioGroup View. The steps for creating the Activities are identical to those in the preceding sections. Therefore you will be provided with the full code of the three main Activity files—AndroidManifest.xml, radiogroup.xml, and testRadioGroup.java. These files are provided for you in the following sections.

AndroidManifest.xml

This section contains the full code of the current AndroidViews' AndroidManifest.xml. If you are following along in Eclipse, modify your Activity's AndroidManifest.xml to look as follows:

```
<?xml version="1.0" encoding="utf-8"?>
<manifest xmlns:android=http://schemas.android.com/apk/res/android
    package="android_programmers_guide.AndroidViews">
    <application android:icon="@drawable/icon">
        <activity android:name=".AndroidViews"
```

```
android:label="@string/app_name">
            <intent-filter>
                <action android:name="android.intent.action.MAIN" />
                <category android:name="android.intent.category.LAUNCHER" />
            </intent-filter>
        </activity>
        <activity android:name=".AutoComplete" android:label="AutoComplete">
            <intent-filter>
                  <action android:name="android.intent.action.MAIN" />
                  <category android:name="android.intent.category.LAUNCHER"/>
      </intent-filter>
        </activity>
         <activity android:name=".testButton" android:label="TestButton">
            <intent-filter>
                  <action android:name="android.intent.action.MAIN" />
                  <category android:name="android.intent.category.LAUNCHER"/>
            </intent-filter>
        </activity>
        <activity android:name=".testCheckBox" android:label="TestCheckBox">
            <intent-filter>
                  <action android:name="android.intent.action.MAIN" />
                  <category android:name="android.intent.category.LAUNCHER"/>
            </intent-filter>
        </activity>
        <activity android:name=".testEditText" android:label="TestEditText">
            <intent-filter>
                  <action android:name="android.intent.action.MAIN" />
                  <category android:name="android.intent.category.LAUNCHER"/>
            </intent-filter>
        </activity>
                <activity android:name=".testRadioGroup" android:label="Test
RadioGroup">
            <intent-filter>
                  <action android:name="android.intent.action.MAIN" />
                  <category android:name="android.intent.category.LAUNCHER"/>
            </intent-filter>
        </activity>
    </application>
</manifest>
```

radiogroup.xml

This section shows the complete code of the radiogroup.xml. Create a new XML file in your project named radiogroup.xml using the instructions outlined earlier in this chapter. Use the following code to model your file.

```
<?xml version="1.0" encoding="utf-8"?>
<LinearLayout xmlns:android=http://schemas.android.com/apk/res/android
```

```
    android:orientation="vertical"
    android:layout_width="fill_parent"
    android:layout_height="fill_parent"
    >
<RadioGroup android:id="@+id/testRadioGroup"
android:layout_width="fill_parent"
android:layout_height="wrap_content" >
        <RadioButton
            android:text="Radio 1"
            android:id="@+id/radio1"
            />
         <RadioButton
            android:text="Radio 2"
            android:id="@+id/radio2" />
 </RadioGroup>
<Button android:id="@+id/enableButton"
android:layout_width="fill_parent"
android:layout_height="wrap_content"
android:text="Set isEnabled"/>
<Button android:id="@+id/backgroundColorButton"
android:layout_width="fill_parent"
android:layout_height="wrap_content"
android:text="Change Background Color"/>
</LinearLayout>
```

testRadioGroup.java

This section covers the final file needed to implement your RadioGroup Activity. Create a new .java file in your project named testRadioGroup.java. This file is the main file of the Activity and contains the actionable code. Use the following code in your testRadioGroup.java to finish this Activity.

```
package android_programmers_guide.AndroidViews;

import android.app.Activity;
import android.os.Bundle;
import android.view.View;
import android.widget.RadioGroup;
import android.widget.Button;
import android.graphics.Color;

public class testRadioGroup extends Activity {
    @Override
    public void onCreate(Bundle icicle) {
```

```
        super.onCreate(icicle);
        setContentView(R.layout.radiogroup);

        final RadioGroup radiogroup = (RadioGroup)
findViewById(R.id.testRadioGroup);

        final Button changeButton = (Button)findViewById(R.id.enableButton);
        changeButton.setOnClickListener(new Button.OnClickListener() {
              public void onClick(View v){
                    changeOption(radiogroup); }
        });
        final Button changeButton2 = (Button)
findViewById(R.id.backgroundColorButton);
        changeButton2.setOnClickListener(new Button.OnClickListener() {
              public void onClick(View v){
                    changeOption2(radiogroup);
              }
        });
    }
    public void changeOption(RadioGroup radiogroup){
      if (radiogroup.isEnabled()){
            radiogroup.setEnabled(false);
      }
      else{
            radiogroup.setEnabled(true);
      }     }
    public void changeOption2(RadioGroup radiogroup){
      radiogroup.setBackgroundColor(Color.RED);
    }
}
```

AndroidViews.java

The last step to create this Activity is to edit the AndroidViews.java. If you want to call the testRadioGroup Activity from the main AndroidViews Activity, you must add code to the AndroidViews.java. Compare the following code with that in your current AndroidViews.java. Add the needed code to complete your file.

```
package android_programmers_guide.AndroidViews;

import android.app.Activity;
import android.os.Bundle;
import android.view.Menu;
import android.content.Intent;

public class AndroidViews extends Activity {
    /** Called when the Activity is first created. */
```

```
    @Override
    public void onCreate(Bundle icicle) {
        super.onCreate(icicle);
        setContentView(R.layout.main);
       }
@Override
public boolean onCreateOptionsMenu(Menu menu) {
 super.onCreateOptionsMenu(menu);

 menu.add(0, 0, "AutoComplete");
 menu.add(0, 1, "Button");
 menu.add(0, 2, "CheckBox");
 menu.add(0, 3, "EditText");
 menu.add(0, 4, "RadioGroup");
 menu.add(0, 5, "Spinner");
 return true;
}
@Override
public boolean onOptionsItemSelected(Menu.Item item){
 switch (item.getId()) {
 case 0:
     showAutoComplete();
     return true;
 case 1:
     showButton();
     return true;
 case 2:
     showCheckBox();
     return true;
 case 3:
     showEditText();
     return true;
 case 4:
     showRadioGroup();
     return true;
 case 5:
     showSpinner();
     return true;
 }
 return true;
}
public void showButton() {
      Intent showButton = new Intent(this, testButton.class);
      startActivity(showButton);
```

```
}
public void showAutoComplete(){
      Intent autocomplete = new Intent(this, AutoComplete.class);
      startActivity(autocomplete);
}
public void showCheckBox(){
      Intent checkbox = new Intent(this, testCheckBox.class);
      startActivity(checkbox);
 }
public void showEditText() {
      Intent edittext = new Intent(this, testEditText.class);
      startActivity(edittext);
 }
public void showRadioGroup(){
      Intent radiogroup = new Intent(this, testRadioGroup.class);
      startActivity(radiogroup);
}
public void showSpinner(){
}
}
```

Launch your application and select the RadioGroup option from the Menu (shown earlier in Figure 8-1).

The following illustration shows what the RadioGroup Activity looks like.

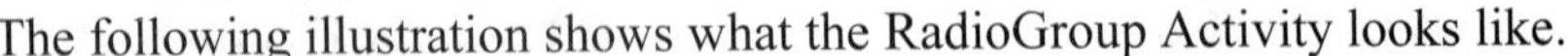

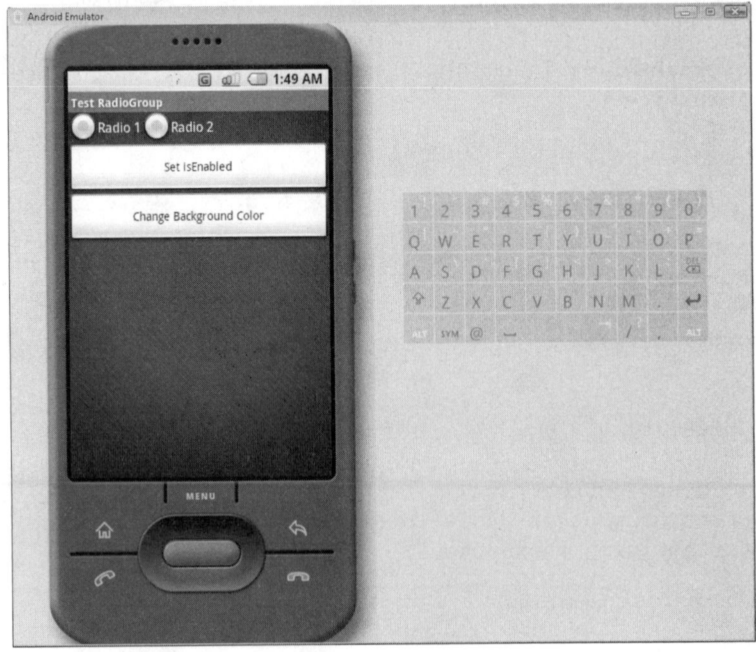

Try clicking the Set isEnabled and Change Background Color Buttons. The results are depicted in the following illustrations. Notice that the Set isEnabled Button for the RadioGroup disables the group, whereas the Change Background Color Button changes the group's background color.

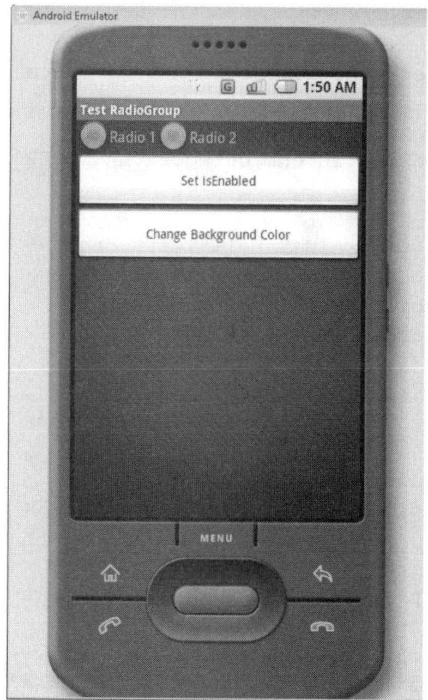

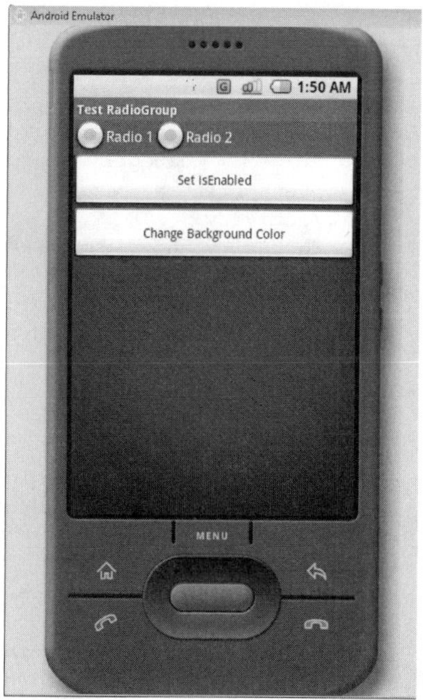

Spinner

In this section you will be creating an Activity for the Spinner View. A Spinner View is similar to a ComboBox in other programming languages. The steps for creating the Activities are identical to those in the preceding sections. Therefore you will be provided with the full code of the three main Activity files—AndroidManifest.xml, spinner.xml, and testSpinner.java. These files are provided for you in the following sections.

AndroidManifest.xml

This section contains the full code of the current AndroidViews' AndroidManifest.xml. If you are following along in Eclipse, modify your Activity's AndroidManifest.xml to look as follows:

```
<?xml version="1.0" encoding="utf-8"?>
<manifest xmlns:android=http://schemas.android.com/apk/res/android
```

```
    package="android_programmers_guide.AndroidViews">
    <application android:icon="@drawable/icon">
        <activity android:name=".AndroidViews"
android:label="@string/app_name">
            <intent-filter>
                <action android:name="android.intent.action.MAIN" />
                <category android:name="android.intent.category.LAUNCHER" />
            </intent-filter>
        </activity>
        <activity android:name=".AutoComplete" android:label="AutoComplete">
            <intent-filter>
                  <action android:name="android.intent.action.MAIN" />
                  <category android:name="android.intent.category.LAUNCHER"/>
      </intent-filter>
        </activity>
         <activity android:name=".testButton" android:label="TestButton">
            <intent-filter>
                  <action android:name="android.intent.action.MAIN" />
                  <category android:name="android.intent.category.LAUNCHER"/>
            </intent-filter>
        </activity>
        <activity android:name=".testCheckBox" android:label="TestCheckBox">
            <intent-filter>
                  <action android:name="android.intent.action.MAIN" />
                  <category android:name="android.intent.category.LAUNCHER"/>
            </intent-filter>
        </activity>
        <activity android:name=".testEditText" android:label="TestEditText">
            <intent-filter>
                  <action android:name="android.intent.action.MAIN" />
                  <category android:name="android.intent.category.LAUNCHER"/>
            </intent-filter>
        </activity>
                <activity android:name=".testRadioGroup" android:label="Test
RadioGroup">
            <intent-filter>
                  <action android:name="android.intent.action.MAIN" />
                  <category android:name="android.intent.category.LAUNCHER"/>
            </intent-filter>
        </activity>
    <activity android:name=".testSpinner" android:label="Test Spinner">
             <intent-filter>
                    <action android:name="android.intent.action.MAIN" />
                    <category android:name="android.intent.category.LAUNCHER" />
             </intent-filter>
        </activity>
    </application>
</manifest>
```

spinner.xml

This section shows the complete code of the spinner.xml. Create a new XML file in your project named spinner.xml using the instructions outlined earlier in this chapter. Use the following code to model your file.

```
<?xml version="1.0" encoding="utf-8"?>
<LinearLayout xmlns:android=http://schemas.android.com/apk/res/android
    android:orientation="vertical"
    android:layout_width="fill_parent"
    android:layout_height="fill_parent"
    >
<Spinner android:id="@+id/testSpinner"
android:layout_width="fill_parent"
android:layout_height="wrap_content"
/>
<Button android:id="@+id/enableButton"
android:layout_width="fill_parent"
android:layout_height="wrap_content"
android:text="Set isEnabled"/>
<Button android:id="@+id/backgroundColorButton"
android:layout_width="fill_parent"
android:layout_height="wrap_content"
android:text="Change Background Color"/>
</LinearLayout>
```

testSpinner.java

This section covers the final file needed to implement your Spinner Activity. Create a new .java file in your project named testSpinner.java. This file is the main file of the Activity and contains the actionable code. Use the following code in your testSpinner.java to finish this Activity.

```
package android_programmers_guide.AndroidViews;

import android.app.Activity;
import android.os.Bundle;
import android.view.View;
import android.widget.ArrayAdapter;
import android.widget.Spinner;
import android.widget.Button;
import android.graphics.Color;
```

```
public class testSpinner extends Activity {
    @Override
    public void onCreate(Bundle icicle) {
        super.onCreate(icicle);
        setContentView(R.layout.spinner);

        final Spinner spinner = (Spinner) findViewById(R.id.testSpinner);
        ArrayAdapter<String> adapter = new ArrayAdapter<String>(this,
                android.R.layout.simple_spinner_item, Months);
adapter.setDropDownViewResource(android.R.layout.simple_spinner_dropdown_item);
        spinner.setAdapter(adapter);

        final Button changeButton = (Button)findViewById(R.id.enableButton);
        changeButton.setOnClickListener(new Button.OnClickListener() {
              public void onClick(View v){
                    changeOption(spinner); }
        });
        final Button changeButton2 = (Button)
findViewById(R.id.backgroundColorButton);
        changeButton2.setOnClickListener(new Button.OnClickListener() {
              public void onClick(View v){
                    changeOption2(spinner);
              }
        });
    }
    static final String[] Months = new String[]{
      "January","February","March","April","May","June","July","August",
      "September","October","November","December"
      };
    public void changeOption(Spinner spinner){
      if (spinner.isEnabled()){
            spinner.setEnabled(false);
      }
      else{
            spinner.setEnabled(true);
      }
    }
    public void changeOption2(Spinner spinner){
      spinner.setBackgroundColor(Color.RED);
    }
}
```

AndroidViews.java

The last step to create this Activity is to edit the AndroidViews.java. If you want to call the testSpinner Activity from the main AndroidViews Activity, you must add code to the

AndroidViews.java. Compare the following code with that in your current AndroidViews.java. Add the needed code to complete your file.

```
package android_programmers_guide.AndroidViews;

import android.app.Activity;
import android.os.Bundle;
import android.view.Menu;
import android.content.Intent;

public class AndroidViews extends Activity {
    /** Called when the Activity is first created. */
    @Override
    public void onCreate(Bundle icicle) {
        super.onCreate(icicle);
        setContentView(R.layout.main);
       }
@Override
public boolean onCreateOptionsMenu(Menu menu) {
 super.onCreateOptionsMenu(menu);

 menu.add(0, 0, "AutoComplete");
 menu.add(0, 1, "Button");
 menu.add(0, 2, "CheckBox");
 menu.add(0, 3, "EditText");
 menu.add(0, 4, "RadioGroup");
 menu.add(0, 5, "Spinner");
 return true;
}
@Override
public boolean onOptionsItemSelected(Menu.Item item){
 switch (item.getId()) {
 case 0:
     showAutoComplete();
     return true;
 case 1:
     showButton();
     return true;
 case 2:
     showCheckBox();
     return true;
```

```
        case 3:
            showEditText();
            return true;
        case 4:
            showRadioGroup();
            return true;
        case 5:
            showSpinner();
            return true;
        }
        return true;
    }
    public void showButton() {
            Intent showButton = new Intent(this, testButton.class);
            startActivity(showButton);
    }
    public void showAutoComplete(){
            Intent autocomplete = new Intent(this, AutoComplete.class);
            startActivity(autocomplete);
    }
    public void showCheckBox(){
            Intent checkbox = new Intent(this, testCheckBox.class);
            startActivity(checkbox);
    }
    public void showEditText() {
            Intent edittext = new Intent(this, testEditText.class);
            startActivity(edittext);
    }
    public void showRadioGroup(){
            Intent radiogroup = new Intent(this, testRadioGroup.class);
            startActivity(radiogroup);
    }
    public void showSpinner(){
            Intent spinner = new Intent(this, testSpinner.class);
            startActivity(spinner);
    }
}
```

Launch your application and select the Spinner option from the Menu (shown earlier in Figure 8-1).

The following illustration shows what the Spinner Activity looks like.

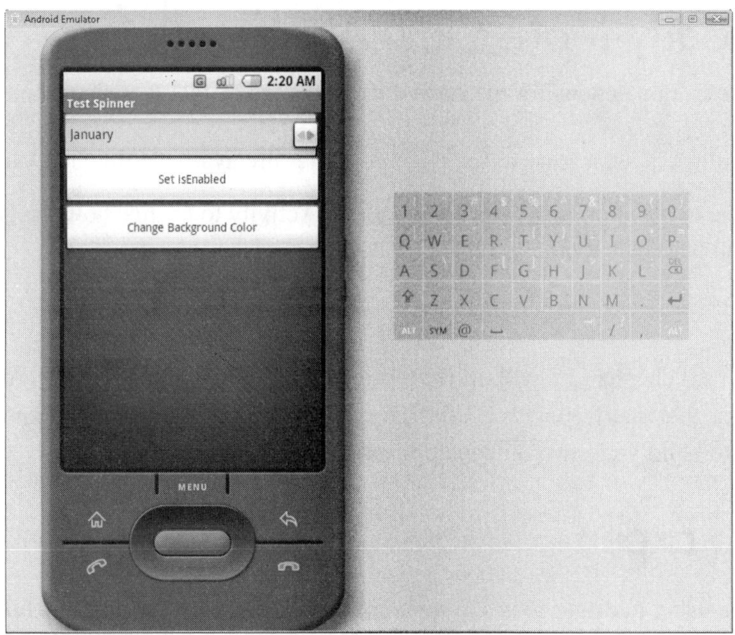

Try clicking the Set isEnabled and Change Background Color Buttons. The results are depicted in the following illustrations.

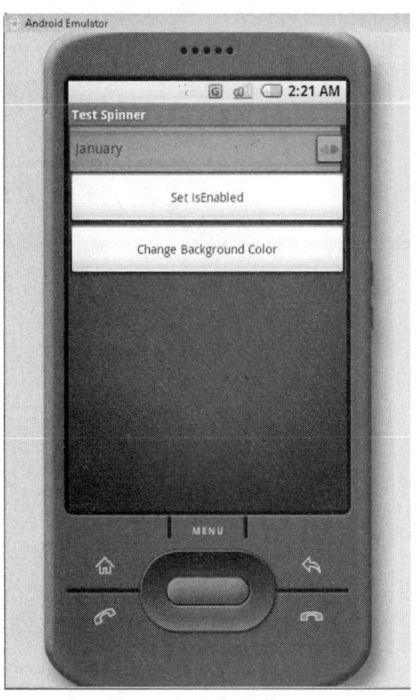

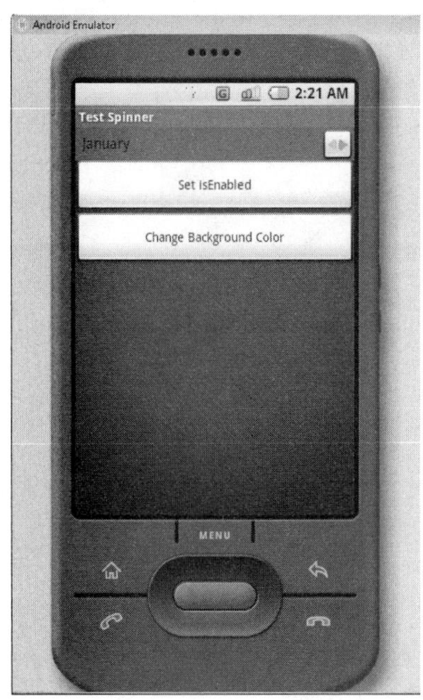

Try This Modify More View Attributes

Modify the Button actions for the Activities to change different attributes on each View:

- Use Eclipse's list feature to see what attributes are available for each View.
- Edit the two Button functions on any given Activity to change how the Buttons interact with that View.

In the next chapter you will utilize more of the Google API. You will create applications that interface with GTalk. This will give you a great base of knowledge on which to build your own unique applications.

Ask the Expert

Q: If I am using multiple Views in my application, can I just import the full widget package using the call *import android.widget.*;*?

A: Yes. However, I would use calls like this sparingly. When you import the entire root of a specific package, you add all the code of that package to your Activity. This can slow down your Activity if not managed. I try to import just those sections of specific packages that I need, in an attempt to reduce the amount of code in the Activity.

Chapter 9

Using the Cell Phone's GPS Functionality

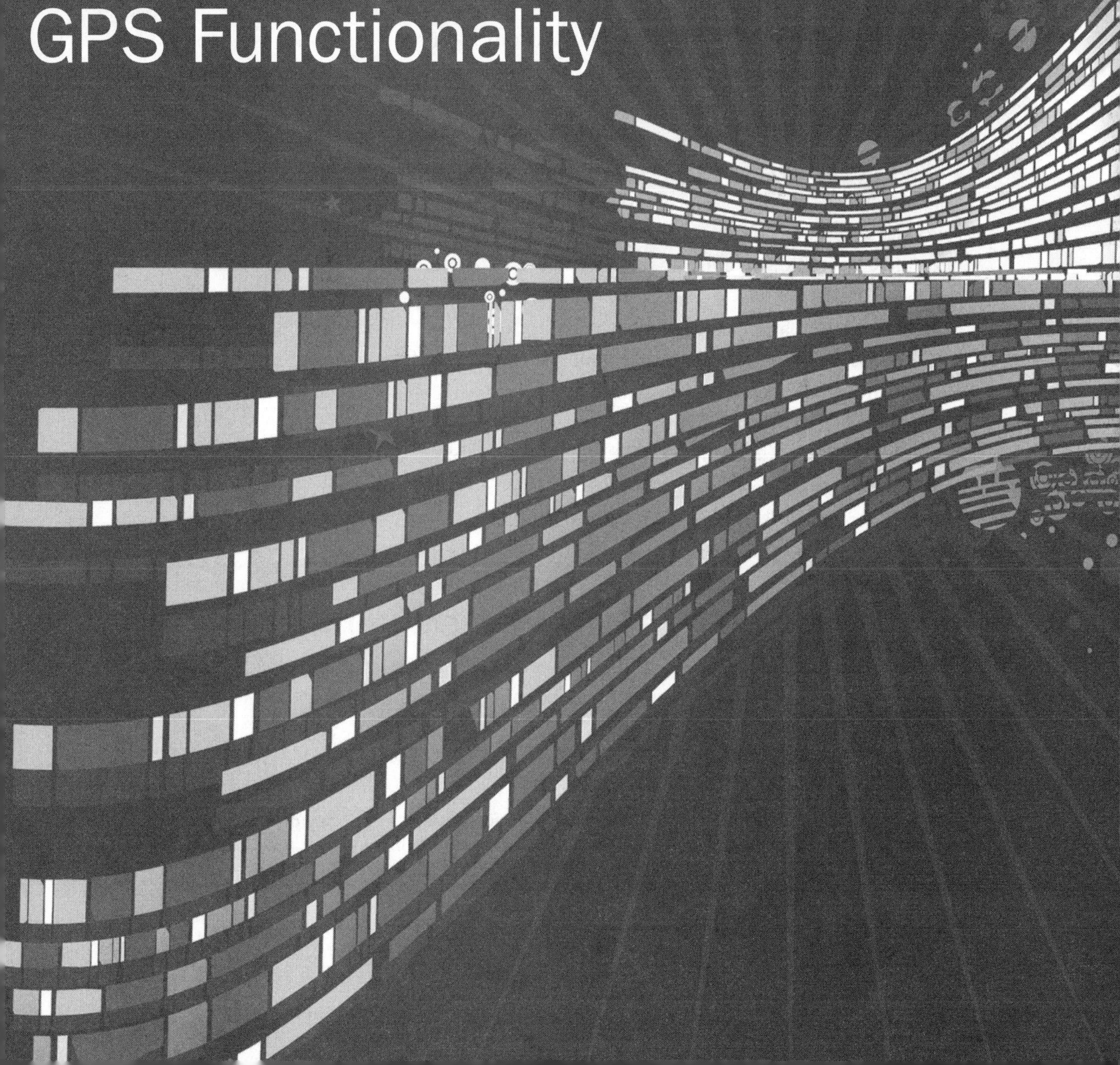

Key Skills & Concepts

- Using the Android location-based service APIs
- Obtaining coordinate data from the GPS hardware
- Changing your Activity's look and feel with a RelativeLayout
- Using a MapView to plot your current location
- Using Google Maps to find your current location

In this chapter, you are going to learn about the Android Location-Based API. This chapter is invaluable if you want to leverage the ability of Android to work with the Global Positioning System (GPS) hardware of a device. You will use the Android Location-Based API to collect your current position and display that location to the screen. Toward the end of this chapter, you will use Google Maps to display your current location on your cell phone.

You will also learn some new techniques that will add some depth and creativity to your Activities. Resources such as RelativeLayouts and small buttons will let you create more user-friendly and visually appealing Activities for Android.

In the first section of this chapter, you will learn about using your device's GPS hardware to obtain your current location. However, before you jump into that section, you need to create your project for this chapter. Create a new Project in Eclipse and name it **AndroidLBS**.

Using the Android Location-Based API

The Android SDK contains an API that is specifically geared to help you interface your Activity with any GPS hardware that may be in your device. This chapter assumes that your device will include GPS hardware.

CAUTION

Just as Android-based cell phones are not required to include a camera, they are not required to include GPS hardware either, although many models likely will include both a camera and GPS hardware. Android included the Android Location-Based API in anticipation that GPS hardware will be included in many cell phones.

Because you are working on a software-based emulator, and not on a real device, the presence of GPS hardware has to be simulated. In this case, Android provides a file in the adb server that simulates having GPS hardware. The file is located at

```
data/misc/location/<provider>
```

where <provider> represents the location information provider. The provider that Android supplies to you is

```
data/misc/location/gps
```

TIP

You can have multiple providers to simulate different scenarios. Therefore, you can create a provider named *test* or *gps1*; whichever you prefer.

Within the specific provider's folder could be any number of files that will hold the sample coordinates that you want Android to use. When you are using the Android Emulator, you can use the following types of files to store/retrieve GPS style coordinates. Each of these file types has a different format for providing information to the Android Location-Based API.

- kml
- nmea
- track

Let's take a look at what each of these files does and how they differ from each other.

Creating a kml File

A .kml file is a Keyhole Markup Language file. These files were originally developed for, and can be created by, Google Earth. The Android Location-Based API can parse a .kml file for coordinates to simulate a GPS.

NOTE

If you do not have Google Earth, it is a free download from Google. Installing it may be worth your time if you want to develop more Android Location-Based API Activities.

To create a .kml file from Google Earth, open Google Earth and navigate to a location. In the following illustration, I have navigated to Tampa, Florida.

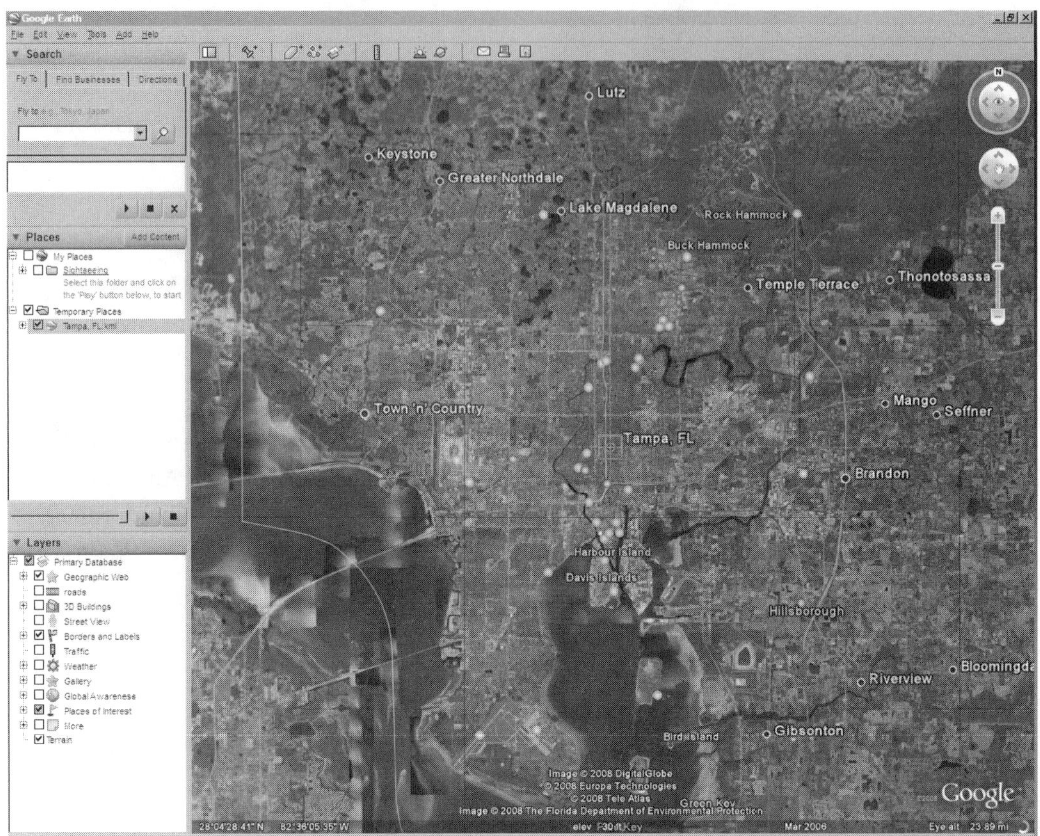

Choose File | Save As and choose KML. In the example, this produces a .kml file with the navigation information for Tampa, Florida. The following .kml code is from this file. Pay close attention to the <coordinates> tag, which is what the Android Location-Based API would be read in.

```
<?xml version="1.0" encoding="UTF-8"?>
<kml xmlns="http://earth.google.com/kml/2.2">
<Document>
<name>Tampa, FL.kml</name>
<Styleid="default+icon=http://maps.google.com/mapfiles/kml/pal3/icon52.png">
     <IconStyle>
```

```
                <scale>1.1</scale>
                <Icon>
            <href>http://maps.google.com/mapfiles/kml/pal3/icon52.png</href>
                </Icon>
        </IconStyle>
        <LabelStyle>
            <scale>1.1</scale>
            </LabelStyle>
        </Style>
        <Styleid="default+icon=http://maps.google.com/mapfiles/kml/pal3/icon60.png">
        <IconStyle>
                <Icon>
            <href>http://maps.google.com/mapfiles/kml/pal3/icon60.png</href>
                </Icon>
            </IconStyle>
        </Style>
        <StyleMapid="default+nicon=http://maps.google.com/mapfiles/kml/pal3/
icon60.png+hicon=http://maps.google.com/mapfiles/kml/pal3/icon52.png">
                <Pair>
                <key>normal</key>
        <styleUrl>#default+icon=http://maps.google.com/mapfiles/kml/pal3/
icon60.png</styleUrl>
            </Pair>
            <Pair>
                <key>highlight</key>
        <styleUrl>#default+icon=http://maps.google.com/mapfiles/kml/pal3/
icon52.png</styleUrl>
            </Pair>
        </StyleMap>
        <Placemark>
            <name>Tampa, FL</name>
            <open>1</open>
            <address>Tampa, FL</address>
            <LookAt>
                <longitude>-82.451142</longitude>
                <latitude>27.98146</latitude>
                <altitude>0</altitude>
                <range>38427.828125</range>
                <tilt>0</tilt>
                <heading>0</heading>
            </LookAt>
<styleUrl>#default+nicon=http://maps.google.com/mapfiles/kml/pal3/
icon60.png+hicon=http://maps.google.com/mapfiles/kml/pal3/icon52.png</styleUrl>
                <Point>
                <coordinates>-82.451142,27.98146,0</coordinates>
            </Point>
</Placemark>
</Document>
</kml>
```

You can create your own .kml files with Google Earth to simulate different locations. This is useful when you want to make an Activity that responds differently depending on the location of the user. The ease of creating .kml files makes this a very flexible alternative for simulating GPS hardware.

What Is a track File?

The file that Android provides in the *gps* folder is an .nmea file (National Marine Electronics Association file). An .nmea file can be output from many popular GPS products. These files are in a common format and can contain multiple coordinates and elevations, representing trips or *tracks*. The following sections discuss getting and opening this file in Windows and Linux, respectively.

Getting the nmea File in Windows

The nmea file provided by Android represents a short trip through San Francisco.

Let's take a look inside the nmea file. Use the adb tool to pull the file from the server to your desktop:

```
adb pull <remote file> <local file>
```

The following illustration depicts the use of the **adb pull** command to retrieve the file.

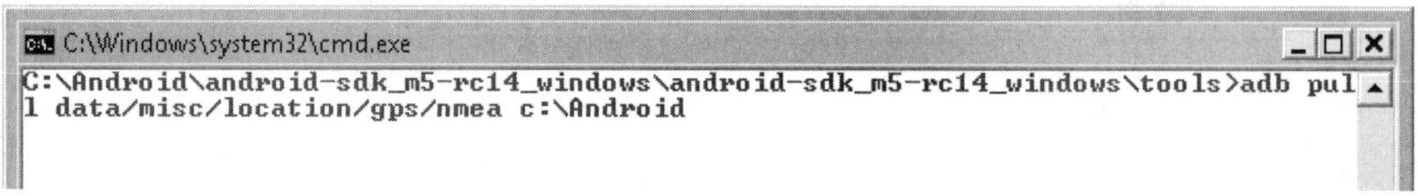

```
C:\Windows\system32\cmd.exe
C:\Android\android-sdk_m5-rc14_windows\android-sdk_m5-rc14_windows\tools>adb pul
l data/misc/location/gps/nmea c:\Android
```

If your command executes successfully, you should see a message like that shown in the following illustration, indicating the size of the file downloaded.

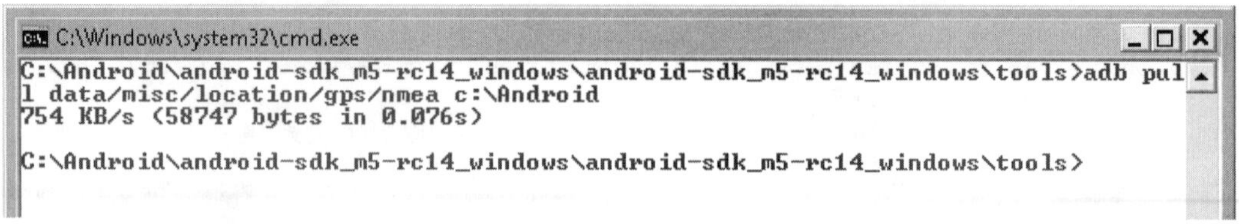

```
C:\Windows\system32\cmd.exe
C:\Android\android-sdk_m5-rc14_windows\android-sdk_m5-rc14_windows\tools>adb pul
l data/misc/location/gps/nmea c:\Android
754 KB/s (58747 bytes in 0.076s)

C:\Android\android-sdk_m5-rc14_windows\android-sdk_m5-rc14_windows\tools>
```

Navigating to the C:\Android folder, you can see that the **adb pull** tool placed the nmea file here (see the following illustration).

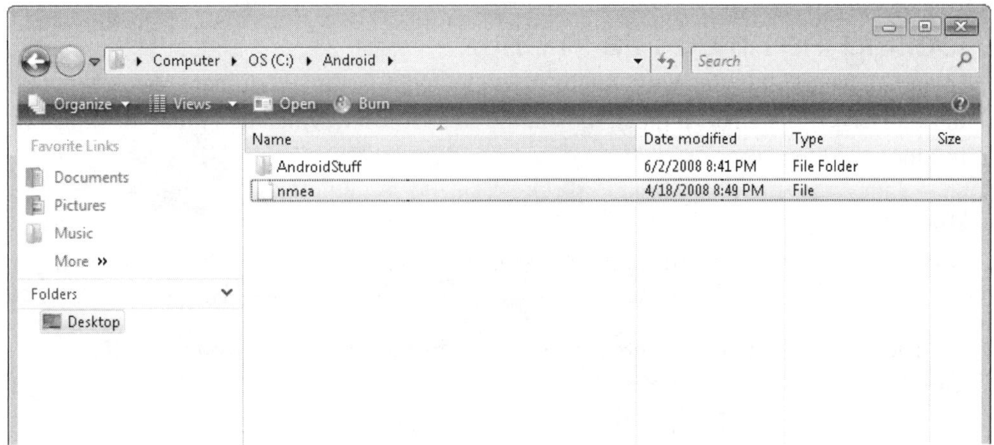

Now that you have the file pulled to your desktop, associate it with Notepad.

Finally, open the file to examine its contents. You should see many rows of coordinate data, as shown here.

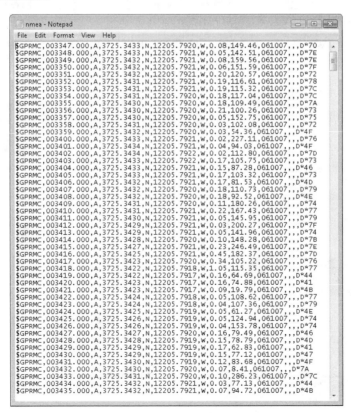

nmea - Notepad

File Edit Format View Help

```
$GPRMC,003347.000,A,3725.3433,N,12205.7920,W,0.08,149.46,061007,,,D*70
$GPRMC,003348.000,A,3725.3433,N,12205.7921,W,0.05,142.51,061007,,,D*7E
$GPRMC,003349.000,A,3725.3432,N,12205.7921,W,0.08,159.56,061007,,,D*7E
$GPRMC,003350.000,A,3725.3432,N,12205.7921,W,0.06,151.59,061007,,,D*7F
$GPRMC,003351.000,A,3725.3432,N,12205.7921,W,0.20,120.57,061007,,,D*72
$GPRMC,003352.000,A,3725.3431,N,12205.7921,W,0.19,116.61,061007,,,D*78
$GPRMC,003353.000,A,3725.3431,N,12205.7921,W,0.19,115.32,061007,,,D*7C
$GPRMC,003354.000,A,3725.3431,N,12205.7920,W,0.18,117.04,061007,,,D*7C
$GPRMC,003355.000,A,3725.3430,N,12205.7920,W,0.18,109.49,061007,,,D*7A
$GPRMC,003356.000,A,3725.3430,N,12205.7920,W,0.21,100.26,061007,,,D*73
$GPRMC,003357.000,A,3725.3430,N,12205.7920,W,0.05,152.75,061007,,,D*75
$GPRMC,003358.000,A,3725.3431,N,12205.7920,W,0.03,102.08,061007,,,D*72
$GPRMC,003359.000,A,3725.3432,N,12205.7920,W,0.03,54.36,061007,,,D*4F
$GPRMC,003400.000,A,3725.3433,N,12205.7921,W,0.02,227.11,061007,,,D*76
$GPRMC,003401.000,A,3725.3434,N,12205.7921,W,0.04,94.03,061007,,,D*4F
$GPRMC,003402.000,A,3725.3434,N,12205.7922,W,0.02,112.80,061007,,,D*7D
$GPRMC,003403.000,A,3725.3433,N,12205.7922,W,0.17,105.75,061007,,,D*73
$GPRMC,003404.000,A,3725.3433,N,12205.7921,W,0.15,87.28,061007,,,D*46
$GPRMC,003405.000,A,3725.3433,N,12205.7921,W,0.17,103.32,061007,,,D*73
$GPRMC,003406.000,A,3725.3432,N,12205.7921,W,0.17,81.53,061007,,,D*4D
$GPRMC,003407.000,A,3725.3432,N,12205.7920,W,0.18,110.73,061007,,,D*79
$GPRMC,003408.000,A,3725.3432,N,12205.7920,W,0.18,92.52,061007,,,D*4E
$GPRMC,003409.000,A,3725.3431,N,12205.7920,W,0.11,180.26,061007,,,D*74
$GPRMC,003410.000,A,3725.3431,N,12205.7921,W,0.22,167.43,061007,,,D*77
$GPRMC,003411.000,A,3725.3430,N,12205.7921,W,0.05,145.95,061007,,,D*79
$GPRMC,003412.000,A,3725.3429,N,12205.7921,W,0.03,200.27,061007,,,D*7F
$GPRMC,003413.000,A,3725.3429,N,12205.7921,W,0.05,141.96,061007,,,D*74
$GPRMC,003414.000,A,3725.3428,N,12205.7920,W,0.10,148.28,061007,,,D*7B
$GPRMC,003415.000,A,3725.3427,N,12205.7921,W,0.23,246.49,061007,,,D*7E
$GPRMC,003416.000,A,3725.3425,N,12205.7921,W,0.45,182.37,061007,,,D*7D
$GPRMC,003417.000,A,3725.3423,N,12205.7920,W,0.34,105.22,061007,,,D*76
$GPRMC,003418.000,A,3725.3422,N,12205.7918,W,1.05,115.35,061007,,,D*77
$GPRMC,003419.000,A,3725.3422,N,12205.7917,W,0.16,64.69,061007,,,D*44
$GPRMC,003420.000,A,3725.3423,N,12205.7917,W,0.16,74.88,061007,,,D*41
$GPRMC,003421.000,A,3725.3423,N,12205.7917,W,0.09,19.79,061007,,,D*4B
$GPRMC,003422.000,A,3725.3424,N,12205.7918,W,0.05,108.62,061007,,,D*77
$GPRMC,003423.000,A,3725.3424,N,12205.7918,W,0.04,107.36,061007,,,D*79
$GPRMC,003424.000,A,3725.3425,N,12205.7919,W,0.05,61.27,061007,,,D*4E
$GPRMC,003425.000,A,3725.3426,N,12205.7919,W,0.05,124.94,061007,,,D*74
$GPRMC,003426.000,A,3725.3426,N,12205.7919,W,0.04,153.78,061007,,,D*74
$GPRMC,003427.000,A,3725.3427,N,12205.7920,W,0.16,79.49,061007,,,D*46
$GPRMC,003428.000,A,3725.3428,N,12205.7919,W,0.15,78.79,061007,,,D*4D
$GPRMC,003429.000,A,3725.3429,N,12205.7919,W,0.17,62.83,061007,,,D*41
$GPRMC,003430.000,A,3725.3429,N,12205.7919,W,0.15,77.12,061007,,,D*47
$GPRMC,003431.000,A,3725.3430,N,12205.7919,W,0.12,83.68,061007,,,D*4F
$GPRMC,003432.000,A,3725.3430,N,12205.7920,W,0.07,8.41,061007,,,D*7A
$GPRMC,003433.000,A,3725.3431,N,12205.7920,W,0.10,286.23,061007,,,D*7C
$GPRMC,003434.000,A,3725.3432,N,12205.7921,W,0.03,77.13,061007,,,D*44
$GPRMC,003435.000,A,3725.3432,N,12205.7921,W,0.07,94.72,061007,,,D*4B
```

Getting the nmea File in Linux

If you are using Linux for your Android development, begin a terminal session to access the adb server. Let's take a quick look at the steps for retrieving and editing the nmea file in Linux.

NOTE

The screenshots in this section were taken in the Fedora distribution of Linux.

The first step is to open a new terminal session (Applications | System Tools | Terminal).

Next, use the **adb pull** command to pull the nmea file to the Android folder:

```
adb pull data/misc/location/gps/nmea Android/
```

If you read the Windows directions for getting the nmea file, you'll notice a slight difference in the syntax for Linux; the inclusion of c:\ is unnecessary because of the difference in directory structures.

After you execute the command from the terminal, the resulting screen should appear as shown in the following illustration.

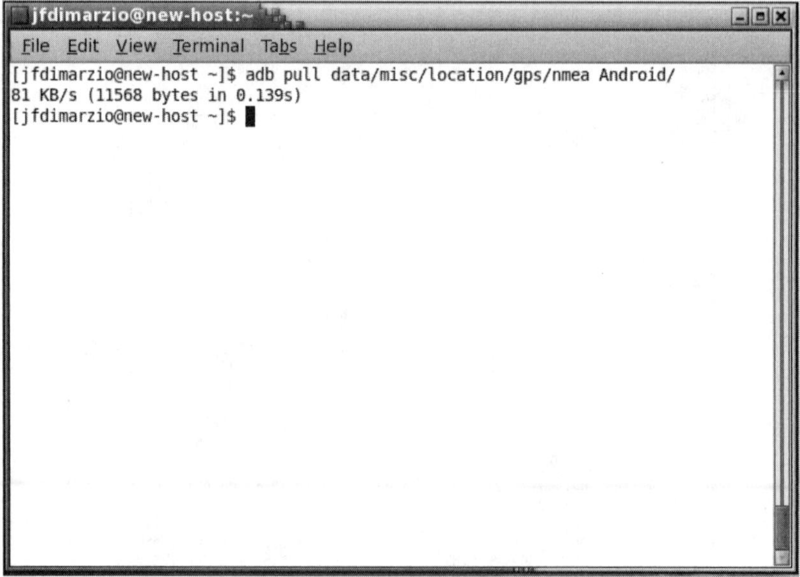

Use the **ls** command to list the files in the Android folder. If the command executed correctly, the nmea file should appear as shown in the following illustration.

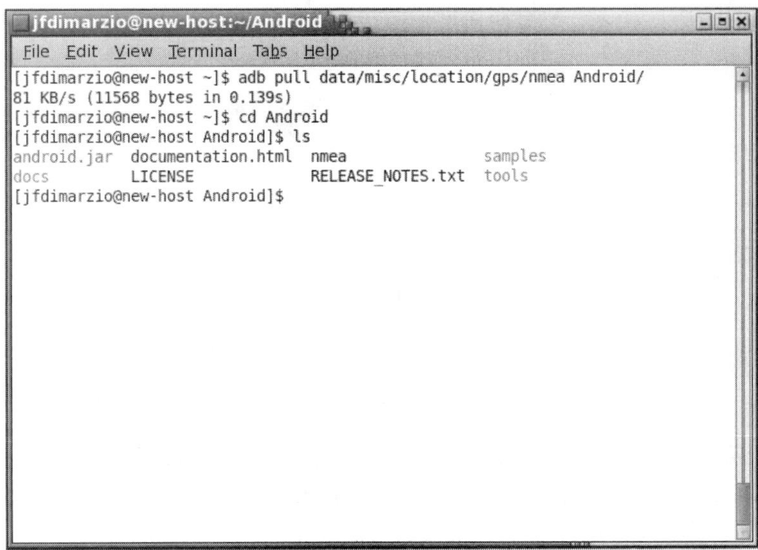

I used the Fedora GUI to navigate to the nmea file and open it in the system Text Editor.

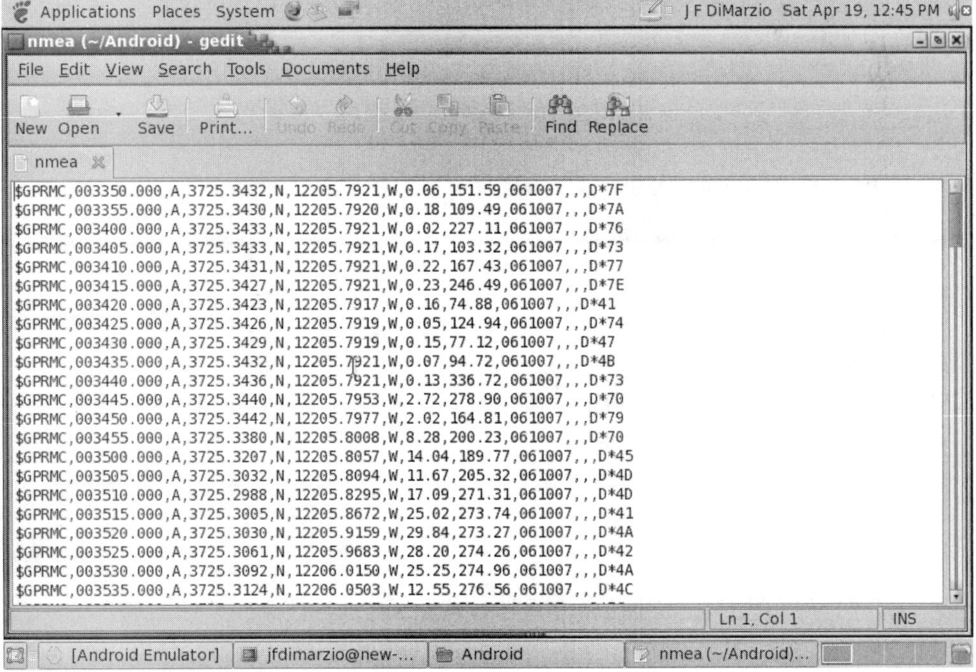

TIP

You could just as easily use the vi editor to open, read, and modify the nmea file from the command line.

Now that you have examined the nmea file and the different methods for simulating a GPS device, you can begin to use the Android Location-Based API to create a full-featured Activity.

Reading the GPS with the Android Location-Based API

The remainder of this chapter is devoted to building an Activity, AndroidLBS, that identifies the location of the user from the nmea file on the server. The first iteration of this Activity will be fairly simple.

You are going to create a simple Activity that will get the user's current GPS location. You can then display that location as a longitude and latitude coordinate pair. In doing this, you will get a good introduction to the Android Location-Based API and how it functions.

Creating the AndroidLBS Activity

The following are the steps for creating this simple Activity:

1. Adjust the permissions level.
2. Create your Activity's layout.
3. Write the code to run your Activity.
4. Run the Activity.

Adjusting the Permissions Level

The first step in working with the Android Location-Based API is to adjust the permissions level. Using the Android Location-Based API itself does not require any specific permission, but using the Android Location-Based API to access location information on the GPS does.

There are two ways you can set the permission from Eclipse. The first is through the Android Manifest Permissions wizard, which you used in Chapter 7. In Eclipse, double-click AndroidManifest.xml to open the Android Manifest Overview window. Click the Permission link and, using the same method described in Chapter 7, add the ACCESS_GPS and ACCESS_LOCATION Uses Permission as shown in the following illustration.

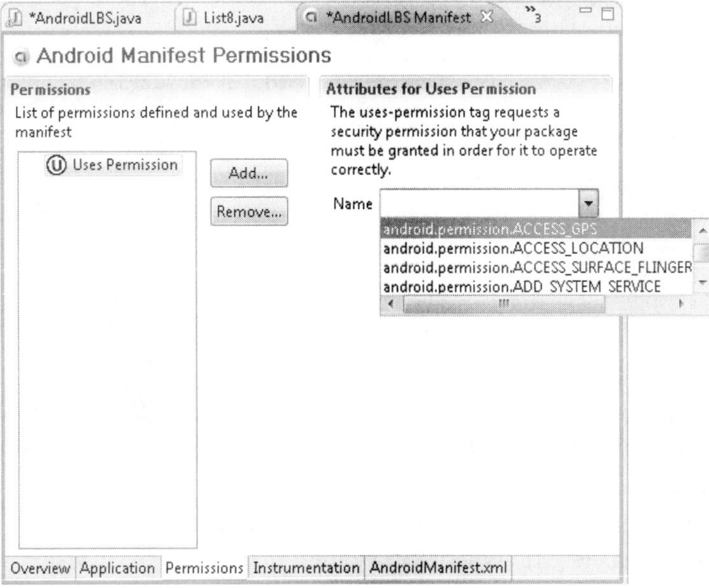

The second way you can add the permission values to your Activity is to edit AndroidManifest.xml manually. You would need to add the following lines to AndroidManifest.xml:

```
<uses-permission android:name="android.permission.ACCESS_GPS">
</uses-permission>
<uses-permission android:name="android.permission.ACCESS_LOCATION">
</uses-permission>
```

The syntax here is to add the permission name within the <uses-permission> tag.

When you have finished adding the permissions, your AndroidManifest.xml file should look like the following code snippet. This code should look pretty familiar by now. You are using just one Activity in the Intent Filter, and a pair of permissions.

```
<?xml version="1.0" encoding="utf-8"?>
<manifest xmlns:android=http://schemas.android.com/apk/res/android
    package="android_programmers_guide.AndroidLBS">
    <application android:icon="@drawable/icon">
        <activity android:name=".AndroidLBS"
android:label="@string/app_name">
            <intent-filter>
                <action android:name="android.intent.action.MAIN" />
                <category android:name="android.intent.category.LAUNCHER" />
            </intent-filter>
        </activity>
    </application>
<uses-permission android:name="android.permission.ACCESS_GPS">
</uses-permission><uses-permission
android:name="android.permission.ACCESS_LOCATION">
</uses-permission></manifest>
```

Creating Your Layout

To begin designing your layout, open main.xml in Eclipse. In total, you will be adding one Button and four TextViews to the layout. The Button will call the information from the GPS and display it to the TextViews.

Set up the Button as follows, which creates a Button layout that fills the top part of the screen and contains the text "Where Am I":

```
<Button
      android:id="@+id/gpsButton"
    android:layout_width="fill_parent"
    android:layout_height="wrap_content"
    android:text="Where Am I"
    />
```

Next, set up the four TextViews. You should arrange them in the layout so that they appear as two TextViews on top of two more TextViews. This will let you use two of them as labels for the others. To accomplish this, you need to implement two more LinearLayouts.

Notice that all the elements in main.xml are contained in a LinearLayout tag. This tag binds the elements within it to certain rules. For LinearLayouts, the elements are stacked one after the other either in a vertical or horizontal orientation.

The orientation of the LinearLayout is governed by the android:orientation attribute. If this attribute is not assigned, the layout defaults to horizontal. Figure 9-1 shows what a vertical LinearLayout does.

Notice that there are several slots or shelves stacked vertically. You can place elements on these shelves to stack your items on the screen. However, if you want to place a few items next to each other on the same shelf of a vertical LinearLayout, then you need to place a horizontal LinearLayout on the shelf first. This concept can be seen in Figure 9-2.

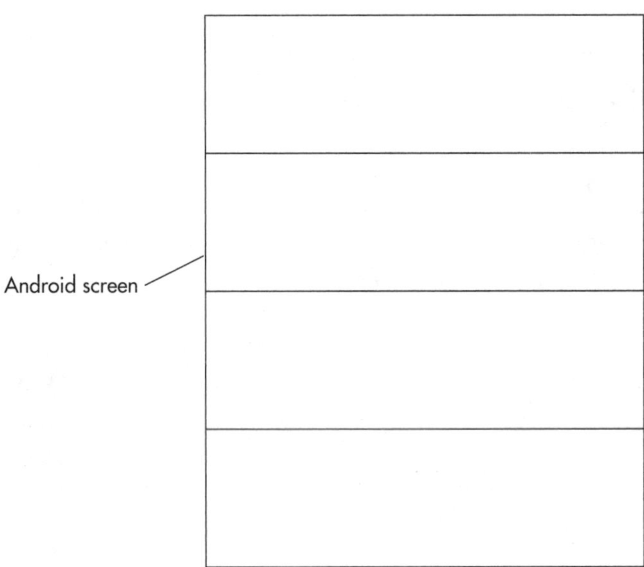

Figure 9-1 Vertical LinearLayout

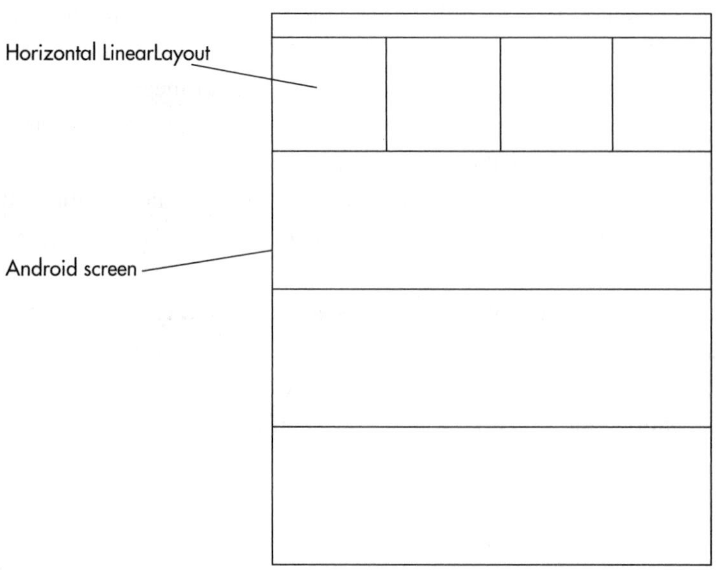

Figure 9-2 Vertical LinearLayout with embedded horizontal LinearLayout

You can now stack elements next to each other and above and below each other. This is the concept you need to employ in this Activity. Therefore, under the button, add a horizontal LinearLayout to hold two of the TextViews.

```
<LinearLayout xmlns:android=http://schemas.android.com/apk/res/android
    android:layout_width="wrap_content"
    android:layout_height="wrap_content"
    >
<TextView
      android:id="@+id/latLabel"
      android:layout_width="wrap_content"
      android:layout_height="wrap_content"
      android:text="Latitude:  "
    />
<TextView
      android:id="@+id/latText"
      android:layout_width="wrap_content"
      android:layout_height="wrap_content"
```

```
    />
 </LinearLayout>
```

These two TextViews hold the label and value for the latitude that you will collect from the GPS. Next, add another horizontal LinearLayout to hold the remaining two TextViews:

```
<LinearLayout xmlns:android=http://schemas.android.com/apk/res/android
    android:layout_width="wrap_content"
    android:layout_height="wrap_content"
    >
 <TextView
      android:id="@+id/lngLabel"
      android:layout_width="wrap_content"
    android:layout_height="wrap_content"
    android:text="Longitude: "
    />
 <TextView
      android:id="@+id/lngText"
      android:layout_width="wrap_content"
    android:layout_height="wrap_content"
    />
</LinearLayout>
```

This will give you a good layout for this particular Activity. Your finished main.xml file should look like this:

```
<?xml version="1.0" encoding="utf-8"?>
<LinearLayout xmlns:android="http://schemas.android.com/apk/res/android"
    android:orientation="vertical"
    android:layout_width="fill_parent"
    android:layout_height="fill_parent"
    >
<Button
    android:id="@+id/gpsButton"
    android:layout_width="fill_parent"
    android:layout_height="wrap_content"
    android:text="Where Am I"
    />
<LinearLayout xmlns:android="http://schemas.android.com/apk/res/android"
    android:layout_width="wrap_content"
    android:layout_height="wrap_content"
    >
```

```
<TextView
      android:id="@+id/latLabel"
      android:layout_width="wrap_content"
    android:layout_height="wrap_content"
    android:text="Latitude: "
    />
<TextView
      android:id="@+id/latText"
      android:layout_width="wrap_content"
    android:layout_height="wrap_content"
    />
 </LinearLayout>
 <LinearLayout xmlns:android="http://schemas.android.com/apk/res/android"
    android:layout_width="wrap_content"
    android:layout_height="wrap_content"
    >
 <TextView
      android:id="@+id/lngLabel"
      android:layout_width="wrap_content"
    android:layout_height="wrap_content"
    android:text="Longitude: "
    />
 <TextView
      android:id="@+id/lngText"
      android:layout_width="wrap_content"
    android:layout_height="wrap_content"
    />
</LinearLayout>
</LinearLayout>
```

Writing the Code to Run Your Activity

Now that you have created your layout, you can begin to write the code that will run your Activity. Your Button needs to call the user's current location from the GPS. Once you have this information, you can then send the longitude and latitude to the corresponding TextViews.

First, you need to add your import statements. The packages that you need to import to complete this Activity include four packages for Views,

```
import android.view.View;
import android.widget.TextView;
import android.content.Context;
import android.widget.Button;
```

and one for the Android Location-Based API:

```
import android.location.LocationManager;
```

Next, create the code for the Button. The goal is to retrieve the current coordinate information from the GPS. You have created a few Buttons already in this book, and the format for this one is no different. You need to set up your Button and load its layout from main.xml. Then you can set up the onClick event to call a function, LoadCoords( ).

```
final Button gpsButton = (Button) findViewById(R.id.gpsButton);
gpsButton.setOnClickListener(new Button.OnClickListener() {
public void onClick(View v){
                    LoadCoords();
            }});
```

The final step to create this Activity is to fill out the code of the LoadCoords( ) function. Create the TextViews that you will post your coordinates to:

```
TextView latText = (TextView) findViewById(R.id.latText);
TextView lngText = (TextView) findViewById(R.id.lngText);
```

NOTE

You do not have to create the two TextViews that you will use as labels because you will not be posting anything to them.

Now create a LocationManager from which you can pull the coordinate values. The important part of this instantiation is that you must pass the LocationManager a context; use the LOCATION_SERVICE:

```
LocationManager myManager =
(LocationManager)getSystemService(Context.LOCATION_SERVICE);
```

To pull the coordinates from myManager, use the getCurrentLocation( ) method. This method needs one parameter, a *provider*, which represents the location that the API will pull the coordinates from. In this case, Android has provided a mock location *gps* that contains the nmea file discussed earlier in this chapter:

```
Double latPoint = myManager.getCurrentLocation("gps").getLatitude();
Double lngPoint = myManager.getCurrentLocation("gps").getLongitude();
```

Finally, take the new Double values and pass them to your TextViews:

```
latText.setText(latPoint.toString());
lngText.setText(lngPoint.toString());
```

Your finished code should look like this:

```
package android_programmers_guide.AndroidLBS;

import android.app.Activity;
import android.os.Bundle;
import android.location.LocationManager;
import android.view.View;
import android.widget.TextView;
import android.content.Context;
import android.widget.Button;


public class AndroidLBS extends Activity {
    /** Called when the activity is first created. */
    @Override
    public void onCreate(Bundle icicle) {
        super.onCreate(icicle);
        setContentView(R.layout.main);
        final Button gpsButton = (Button) findViewById(R.id.gpsButton);

       gpsButton.setOnClickListener(new Button.OnClickListener() {
              public void onClick(View v){
                    LoadCoords();
              }});
    }

    public void LoadCoords(){
      TextView latText = (TextView) findViewById(R.id.latText);
      TextView lngText = (TextView) findViewById(R.id.lngText);
      LocationManager myManager = (LocationManager)
getSystemService(Context.LOCATION_SERVICE);
      Double latPoint = myManager.getCurrentLocation("gps").getLatitude();
      Double lngPoint = myManager.getCurrentLocation("gps").getLongitude();
      latText.setText(latPoint.toString());
      lngText.setText(lngPoint.toString());
    }
    }
```

Running the Activity

Run your Activity in the Android Emulator. The Activity should open to the screen as shown in the following illustration.

Click the Where Am I button. You should see the coordinates shown in this image.

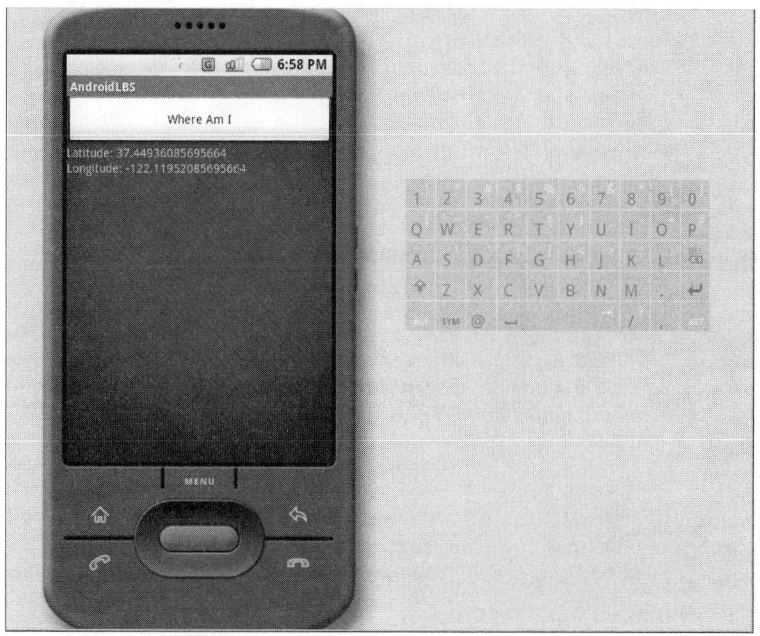

Passing Coordinates to Google Maps

In this section, you will build on the Activity you created in the previous section. The major modification you will make to your AndroidLBS Activity is to pass the coordinates to Google Maps. You will use Google Maps to display the user's current location.

The only change you need to make to your main.xml file is to add a layout for the MapView. In the current version of the Android SDK, the MapView is established as a generic View. Perhaps in a future release there will be a MapView that corresponds to this layout.

```
<view class="com.google.android.maps.MapView"
        android:id="@+id/myMap"
        android:layout_width="wrap_content"
        android:layout_height="wrap_content"/>
```

The complete main.xml file should look like this:

```
<?xml version="1.0" encoding="utf-8"?>
<LinearLayout xmlns:android=http://schemas.android.com/apk/res/android
    android:orientation="vertical"
    android:layout_width="fill_parent"
    android:layout_height="fill_parent"
    >
<Button
      android:id="@+id/gpsButton"
    android:layout_width="fill_parent"
    android:layout_height="wrap_content"
    android:text="Where Am I"
    />

<LinearLayout xmlns:android=http://schemas.android.com/apk/res/android
    android:layout_width="wrap_content"
    android:layout_height="wrap_content"
    >
<TextView
      android:id="@+id/latLabel"
      android:layout_width="wrap_content"
    android:layout_height="wrap_content"
    android:text="Latitude: "
    />
<TextView
      android:id="@+id/latText"
      android:layout_width="wrap_content"
    android:layout_height="wrap_content"
    />
```

```
 </LinearLayout>
 <LinearLayout xmlns:android=http://schemas.android.com/apk/res/android
    android:layout_width="wrap_content"
    android:layout_height="wrap_content"
    >
 <TextView
      android:id="@+id/lngLabel"
      android:layout_width="wrap_content"
    android:layout_height="wrap_content"
    android:text="Longitude: "
    />
 <TextView
      android:id="@+id/lngText"
      android:layout_width="wrap_content"
    android:layout_height="wrap_content"
    />
</LinearLayout>

<view class="com.google.android.maps.MapView"
        android:id="@+id/myMap"
        android:layout_width="wrap_content"
        android:layout_height="wrap_content"/>

</LinearLayout>
```

Because you are embedding the MapView within this Activity, you need to change the definition of your class. Currently, your main class extends Activity. However, to properly work with the Google MapView, you must extend MapActivity. Therefore, you need to import the MapActivity package and replace the Activity package with it in your header.

Import the following packages:

```
import com.google.android.maps.MapActivity;
import com.google.android.maps.MapView;
import com.google.android.maps.Point;
import com.google.android.maps.MapController;
```

The Point package will be used to hold point values that represent map coordinates, whereas the MapController will center the map to your Point. These two packages are critical for using the MapView.

Now you are ready to add the code that will establish the map and pass your coordinates to it. First, set up a MapView and assign it the layout from main.xml:

```
MapView myMap = (MapView) findViewById(R.id.myMap);
```

Next, set up a Point and assign it the latPoint and lngPoint values that you retrieved from the GPS:

```
Point myLocation = new Point(latPoint.intValue(),lngPoint.intValue());
```

Now you can create your MapController, which will be used to move the focus of the Google Map to the location you just defined in the Point. Use the getController( ) method from the MapView to establish a controller in your specific Map:

```
MapController myMapController = myMap.getController();
```

The only job left is to use the controller to move the map to your location (to make the map a little more recognizable, set the zoom to 9):

```
myMapController.centerMapTo(myLocation, false);
myMapController.zoomTo(9);
```

What you have just written is all the code needed to utilize Google Maps from your Activity. The full class should look like this:

```
package android_programmers_guide.AndroidLBS;

import android.os.Bundle;
import android.location.LocationManager;
import android.view.View;
import android.widget.TextView;
import android.content.Context;
import android.widget.Button;
import com.google.android.maps.MapActivity;
import com.google.android.maps.MapView;
import com.google.android.maps.Point;
import com.google.android.maps.MapController;

public class AndroidLBS extends MapActivity {
    /** Called when the activity is first created. */
    @Override
    public void onCreate(Bundle icicle) {
        super.onCreate(icicle);
        setContentView(R.layout.main);
        final Button gpsButton = (Button) findViewById(R.id.gpsButton);
       gpsButton.setOnClickListener(new Button.OnClickListener() {
              public void onClick(View v){
                     LoadProviders();
              }});
    }
```

```
    public void LoadProviders(){
      TextView latText = (TextView) findViewById(R.id.latText);
      TextView lngText = (TextView) findViewById(R.id.lngText);
      LocationManager myManager = (LocationManager)
getSystemService(Context.LOCATION_SERVICE);

      Double latPoint =
myManager.getCurrentLocation("gps").getLatitude()*1E6;
      Double lngPoint =
myManager.getCurrentLocation("gps").getLongitude()*1E6;

      latText.setText(latPoint.toString());
      lngText.setText(lngPoint.toString());

      MapView myMap = (MapView) findViewById(R.id.myMap);
      Point myLocation = new Point(latPoint.intValue(),lngPoint.intValue());

      MapController myMapController = myMap.getController();
      myMapController.centerMapTo(myLocation, false);
      myMapController.zoomTo(9);

    }
    }
```

Run the Activity in the Emulator. The Activity should open to a *blank* map, as shown in the following illustration.

Click the Where Am I button and you should see the map focus to, and zoom in on, San Francisco. Take a look at the following illustration to see how your map should appear.

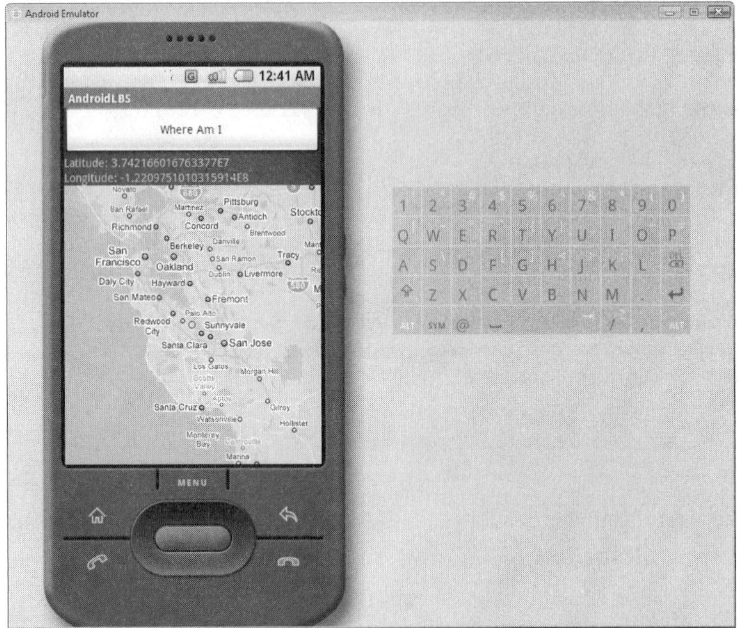

Adding Zoom Controls

For your last exercise in this chapter, you will add two more buttons to your AndroidLBS Activity. These buttons will control the zoom in and zoom out methods of the Google MapView. What makes this modification a little different is that I will introduce a new type of layout for your main.xml file: the RelativeLayout. Whereas LinearLayouts allow you to place Views directly, one after the other, RelativeLayouts let you place Views on top of each other.

For this Activity, you will be placing the two new buttons over the Google Map. To achieve this effect, place the MapView within the RelativeLayout. With the MapView in the RelativeLayout, you can add the buttons that will be placed over the map.

```
<RelativeLayout xmlns:android="http://schemas.android.com/apk/res/android"
    android:orientation="vertical"
    android:layout_width="fill_parent"
```

```
    android:layout_height="fill_parent"
    >
  <view class="com.google.android.maps.MapView"
        android:id="@+id/myMap"
        android:layout_width="wrap_content"
        android:layout_height="wrap_content"/>
</RelativeLayout>
```

Now you can add your two additional buttons. Place the buttons so that they appear in the upper-left and lower-left corners of the MapView. You need to make one change to the standard Button layout. By default, the RelativeLayout adds the Button to align with the top edge of the anchor view, in this case, the MapView. Therefore, in the layout, use the android:layout_alignBottom attribute and assign it the id of the MapView. This will align the button to the bottom of the map.

```
<Button android:id="@+id/buttonZoomIn"
     style="?android:attr/buttonStyleSmall"
            android:text="+"
            android:layout_width="wrap_content"
            android:layout_height="wrap_content" />
            <Button android:id="@+id/buttonZoomOut"
            style="?android:attr/buttonStyleSmall"
            android:text="-"
            android:layout_alignBottom="@+id/myMap"
            android:layout_width="wrap_content"
            android:layout_height="wrap_content" />
```

TIP

Take a close look at the layout attributes for the Button layout. I use a new attribute, *style*, to make this Button a small button.

Your full main.xml file should look like this:

```
<?xml version="1.0" encoding="utf-8"?>
<LinearLayout xmlns:android="http://schemas.android.com/apk/res/android"
    android:orientation="vertical"
    android:layout_width="fill_parent"
    android:layout_height="fill_parent"
    >
<Button
     android:id="@+id/gpsButton"
    android:layout_width="fill_parent"
    android:layout_height="wrap_content"
    android:text="Where Am I"
    />
```

```
<LinearLayout xmlns:android="http://schemas.android.com/apk/res/android"
    android:layout_width="wrap_content"
    android:layout_height="wrap_content"
    >
<TextView
      android:id="@+id/latLabel"
      android:layout_width="wrap_content"
    android:layout_height="wrap_content"
    android:text="Latitude: "
    />
<TextView
      android:id="@+id/latText"
      android:layout_width="wrap_content"
    android:layout_height="wrap_content"
    />
 </LinearLayout>
 <LinearLayout xmlns:android="http://schemas.android.com/apk/res/android"
    android:layout_width="wrap_content"
    android:layout_height="wrap_content"
    >
 <TextView
      android:id="@+id/lngLabel"
      android:layout_width="wrap_content"
    android:layout_height="wrap_content"
    android:text="Longitude: "
    />
 <TextView
      android:id="@+id/lngText"
      android:layout_width="wrap_content"
    android:layout_height="wrap_content"
    />
</LinearLayout>

<RelativeLayout xmlns:android="http://schemas.android.com/apk/res/android"
    android:orientation="vertical"
    android:layout_width="fill_parent"
    android:layout_height="fill_parent"
    >
  <view class="com.google.android.maps.MapView"
        android:id="@+id/myMap"
        android:layout_width="wrap_content"
        android:layout_height="wrap_content"/>
          <Button android:id="@+id/buttonZoomIn"
            style="?android:attr/buttonStyleSmall"
            android:text="+"
            android:layout_width="wrap_content"
            android:layout_height="wrap_content" />
            <Button android:id="@+id/buttonZoomOut"
            style="?android:attr/buttonStyleSmall"
            android:text="-"
```

```
            android:layout_alignBottom="@+id/myMap"
            android:layout_width="wrap_content"
            android:layout_height="wrap_content" />
</RelativeLayout>
</LinearLayout>
```

You are going to make a few modifications to the code. Aside from adding the code for the new views, you need to move some existing code around. To make your Activity more flexible, you need to move the instantiations of the MapView and MapController to the main part of the class. This will allow you to then pass those items into other functions as needed (like those you will create for the zoom in and zoom out features).

```
final MapView myMap = (MapView) findViewById(R.id.myMap);
final MapController myMapController = myMap.getController();
```

Now you can create the code for the two new buttons. Create the buttons as you have done in the past, adding calls to functions you will build next:

```
final Button zoomIn = (Button) findViewById(R.id.buttonZoomIn);
    zoomIn.setOnClickListener(new Button.OnClickListener() {
                public void onClick(View v){
                        ZoomIn(myMap,myMapController);
                }});
final Button zoomOut = (Button) findViewById(R.id.buttonZoomOut);
    zoomOut.setOnClickListener(new Button.OnClickListener() {
        public void onClick(View v){
                        ZoomOut(myMap,myMapController);
                }});
```

Finally, create the functions that will control the zoom in and zoom out feature. The maximum zoom in level is 21 and the minimum is 1. Therefore, in your function, test for the current level before adjusting. This will ensure that you do not run into any problems.

```
    public void ZoomIn(MapView mv, MapController mc){
      if(mv.getZoomLevel()!=21){
      mc.zoomTo(mv.getZoomLevel()+ 1);
      }
    }
    public void ZoomOut(MapView mv, MapController mc){
      if(mv.getZoomLevel()!=1){
            mc.zoomTo(mv.getZoomLevel()- 1);
            }
    }
```

Notice that you pass the MapView and MapController into the functions. From there, it is simply Integer manipulation to set the zoom level. The only tricky part of this function is that the MapController physically moves the MapView to the desired zoom level, whereas the MapView itself holds the zoom value.

TIP

Think of this relationship as being similar to that between a remote control and a television. The remote control tunes the TV to channel 5, but the channel itself is stored on the TV.

Your completed AndroidLBS.java file should look like this:

```
package android_programmers_guide.AndroidLBS;

import android.os.Bundle;
import android.location.LocationManager;
import android.view.View;
import android.widget.TextView;
import android.content.Context;
import android.widget.Button;
import com.google.android.maps.MapActivity;
import com.google.android.maps.MapView;
import com.google.android.maps.Point;
import com.google.android.maps.MapController;

public class AndroidLBS extends MapActivity {
    /** Called when the activity is first created. */
    @Override
    public void onCreate(Bundle icicle) {
        super.onCreate(icicle);
        setContentView(R.layout.main);
        final MapView myMap = (MapView) findViewById(R.id.myMap);
        final MapController myMapController = myMap.getController();
        final Button zoomIn = (Button) findViewById(R.id.buttonZoomIn);
        zoomIn.setOnClickListener(new Button.OnClickListener() {
              public void onClick(View v){
                    ZoomIn(myMap,myMapController);
              }});
        final Button zoomOut = (Button) findViewById(R.id.buttonZoomOut);
        zoomOut.setOnClickListener(new Button.OnClickListener() {
              public void onClick(View v){
                    ZoomOut(myMap,myMapController);
              }});
        final Button gpsButton = (Button) findViewById(R.id.gpsButton);
       gpsButton.setOnClickListener(new Button.OnClickListener() {
             public void onClick(View v){
                   LoadProviders(myMap,myMapController);
             }});


    }
```

```
    public void LoadProviders(MapView mv, MapController mc){
      TextView latText = (TextView) findViewById(R.id.latText);
      TextView lngText = (TextView) findViewById(R.id.lngText);
      LocationManager myManager = (LocationManager)
getSystemService(Context.LOCATION_SERVICE);
      Double latPoint = myManager.getCurrentLocation("gps").getLatitude()*1E6;
      Double lngPoint =
myManager.getCurrentLocation("gps").getLongitude()*1E6;
      latText.setText(latPoint.toString());
      lngText.setText(lngPoint.toString());
      Point myLocation = new Point(latPoint.intValue(),lngPoint.intValue());
      mc.centerMapTo(myLocation, false);
      mc.zoomTo(9);
    }
    public void ZoomIn(MapView mv, MapController mc){
      if(mv.getZoomLevel()!=21){
      mc.zoomTo(mv.getZoomLevel()+ 1);
      }
    }
    public void ZoomOut(MapView mv, MapController mc){
      if(mv.getZoomLevel()!=1){
            mc.zoomTo(mv.getZoomLevel()- 1);
            }
    }
    }
```

Run this Activity in your Android Emulator. The Activity should open to a reset MapView, with the buttons placed as shown in the following illustration.

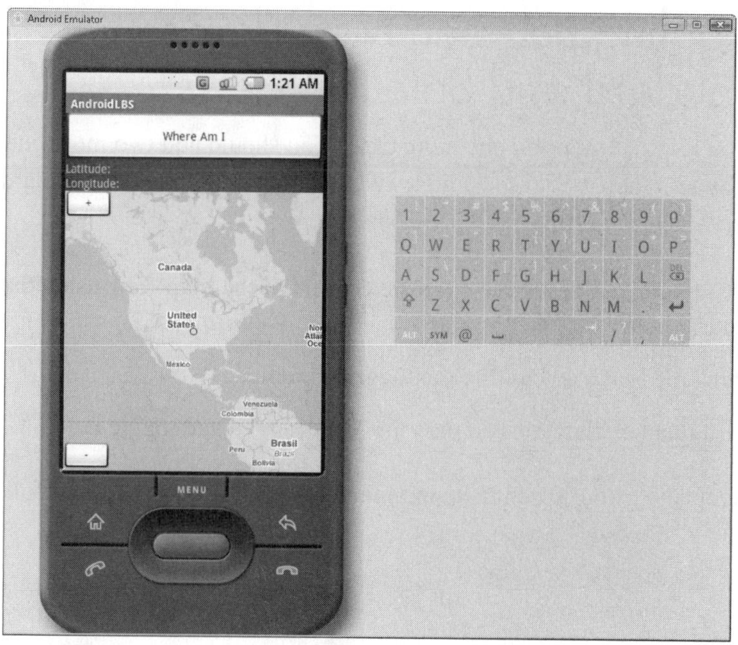

Test the zoom in and zoom out buttons. When you zoom in, you should see something that looks similar to the following illustration.

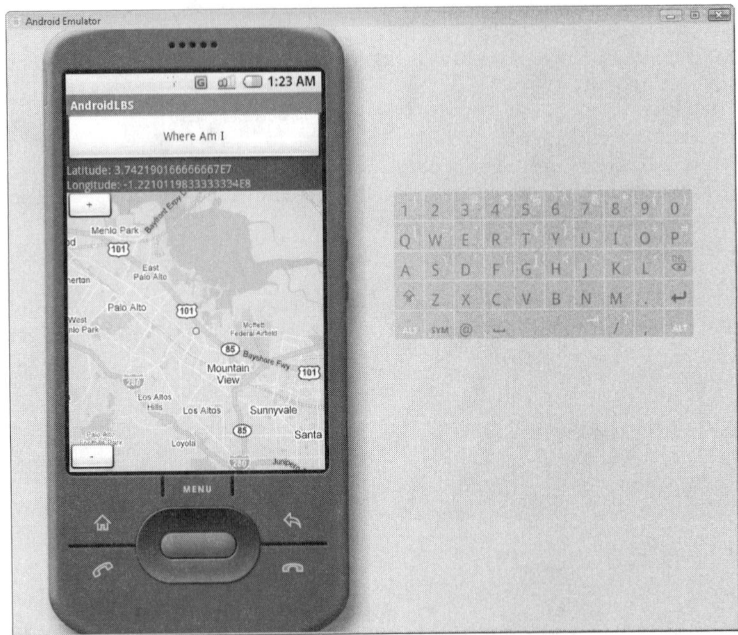

Try This Toggling Between MapView's Standard and Satellite Views

Edit the AndroidLBS Activity one more time. You should add two more buttons to the RelativeLayout. These buttons should toggle the MapView between standard view and satellite view. Here are some points to consider:

- Add the toggle buttons to the opposite corners of the MapView using the align layout attributes.
- Research the MapView to find the toggling method.
- Create a function that you can pass the MapView to and toggle it.

The complete text of solution main.xml and AndroidLBS.java are as follows.

main.xml

```
<?xml version="1.0" encoding="utf-8"?>
<LinearLayout xmlns:android="http://schemas.android.com/apk/res/android"
    android:orientation="vertical"
    android:layout_width="fill_parent"
    android:layout_height="fill_parent"
    >
<Button
      android:id="@+id/gpsButton"
    android:layout_width="fill_parent"
    android:layout_height="wrap_content"
    android:text="Where Am I"
    />

<LinearLayout xmlns:android="http://schemas.android.com/apk/res/android"
    android:layout_width="wrap_content"
    android:layout_height="wrap_content"
    >
<TextView
      android:id="@+id/latLabel"
      android:layout_width="wrap_content"
    android:layout_height="wrap_content"
    android:text="Latitude: "
    />
<TextView
      android:id="@+id/latText"
      android:layout_width="wrap_content"
    android:layout_height="wrap_content"
    />
 </LinearLayout>
 <LinearLayout xmlns:android="http://schemas.android.com/apk/res/android"
    android:layout_width="wrap_content"
    android:layout_height="wrap_content"
    >
 <TextView
      android:id="@+id/lngLabel"
      android:layout_width="wrap_content"
    android:layout_height="wrap_content"
    android:text="Longitude: "
    />
 <TextView
      android:id="@+id/lngText"
      android:layout_width="wrap_content"
    android:layout_height="wrap_content"
    />
</LinearLayout>
```

```
<RelativeLayout xmlns:android="http://schemas.android.com/apk/res/android"
    android:orientation="vertical"
    android:layout_width="fill_parent"
    android:layout_height="fill_parent"
    >
  <view class="com.google.android.maps.MapView"
        android:id="@+id/myMap"
        android:layout_width="wrap_content"
        android:layout_height="wrap_content"/>
          <Button android:id="@+id/buttonZoomIn"
            style="?android:attr/buttonStyleSmall"
            android:text="+"
            android:layout_width="wrap_content"
            android:layout_height="wrap_content" />
          <Button android:id="@+id/buttonMapView"
            style="?android:attr/buttonStyleSmall"
            android:text="Map"
            android:layout_alignRight="@+id/myMap"
            android:layout_width="wrap_content"
            android:layout_height="wrap_content" />
          <Button android:id="@+id/buttonSatView"
            style="?android:attr/buttonStyleSmall"
            android:text="Sat"
            android:layout_alignRight="@+id/myMap"
            android:layout_alignBottom="@+id/myMap"
            android:layout_width="wrap_content"
            android:layout_height="wrap_content" />
          <Button android:id="@+id/buttonZoomOut"
            style="?android:attr/buttonStyleSmall"
            android:text="-"
            android:layout_alignBottom="@+id/myMap"
            android:layout_width="wrap_content"
            android:layout_height="wrap_content" />
</RelativeLayout>
</LinearLayout>
```

AndroidLBS.java

```
package android_programmers_guide.AndroidLBS;

import android.os.Bundle;
import android.location.LocationManager;
import android.view.View;
import android.widget.TextView;
import android.content.Context;
import android.widget.Button;
import com.google.android.maps.MapActivity;
import com.google.android.maps.MapView;
import com.google.android.maps.Point;
import com.google.android.maps.MapController;
```

```
public class AndroidLBS extends MapActivity {
    /** Called when the activity is first created. */
    @Override
    public void onCreate(Bundle icicle) {
        super.onCreate(icicle);
        setContentView(R.layout.main);
        final MapView myMap = (MapView) findViewById(R.id.myMap);
        final MapController myMapController = myMap.getController();
        final Button zoomIn = (Button) findViewById(R.id.buttonZoomIn);
        zoomIn.setOnClickListener(new Button.OnClickListener() {
               public void onClick(View v){
                     ZoomIn(myMap,myMapController);
               }});
        final Button zoomOut = (Button) findViewById(R.id.buttonZoomOut);
        zoomOut.setOnClickListener(new Button.OnClickListener() {
               public void onClick(View v){
                     ZoomOut(myMap,myMapController);
               }});
        final Button gpsButton = (Button) findViewById(R.id.gpsButton);
       gpsButton.setOnClickListener(new Button.OnClickListener() {
              public void onClick(View v){
                    LoadProviders(myMap,myMapController);
              }});
       final Button viewMap = (Button) findViewById(R.id.buttonMapView);
       viewMap.setOnClickListener(new Button.OnClickListener() {
              public void onClick(View v){
                    ShowMap(myMap);
              }});
       final Button viewSat = (Button) findViewById(R.id.buttonSatView);
       viewSat.setOnClickListener(new Button.OnClickListener() {
              public void onClick(View v){
                    ShowSat(myMap);
              }});

    }

    public void LoadProviders(MapView mv, MapController mc){
      TextView latText = (TextView) findViewById(R.id.latText);
      TextView lngText = (TextView) findViewById(R.id.lngText);
      LocationManager myManager = (LocationManager)
getSystemService(Context.LOCATION_SERVICE);
      Double latPoint =
myManager.getCurrentLocation("gps").getLatitude()*1E6;
      Double lngPoint =
myManager.getCurrentLocation("gps").getLongitude()*1E6;
      latText.setText(latPoint.toString());
      lngText.setText(lngPoint.toString());
      Point myLocation = new Point(latPoint.intValue(),lngPoint.intValue());
      mc.centerMapTo(myLocation, false);
      mc.zoomTo(9);
    }
```

```
public void ZoomIn(MapView mv, MapController mc){
  if(mv.getZoomLevel()!=21){
  mc.zoomTo(mv.getZoomLevel()+ 1);
  }
}
public void ZoomOut(MapView mv, MapController mc){
  if(mv.getZoomLevel()!=1){
        mc.zoomTo(mv.getZoomLevel()- 1);
        }
}
public void ShowMap(MapView mv){
        if (mv.isSatellite()){
                mv.toggleSatellite();
        }
}
public void ShowSat(MapView mv){
        if (!mv.isSatellite()){
                mv.toggleSatellite();
        }
}
}
```

When you run your Activity, you should be able to toggle the satellite view on and off, as shown in the following illustrations.

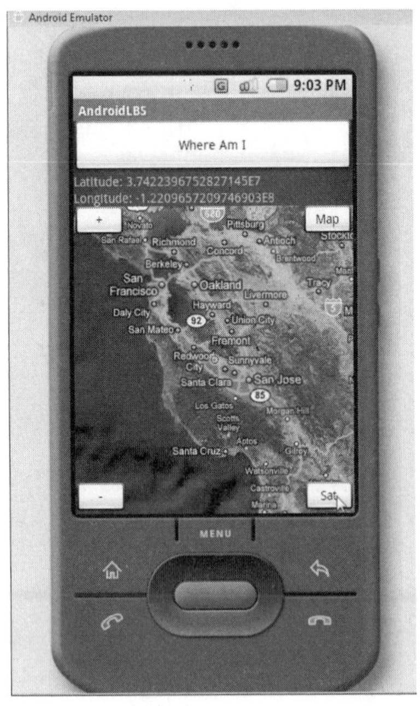

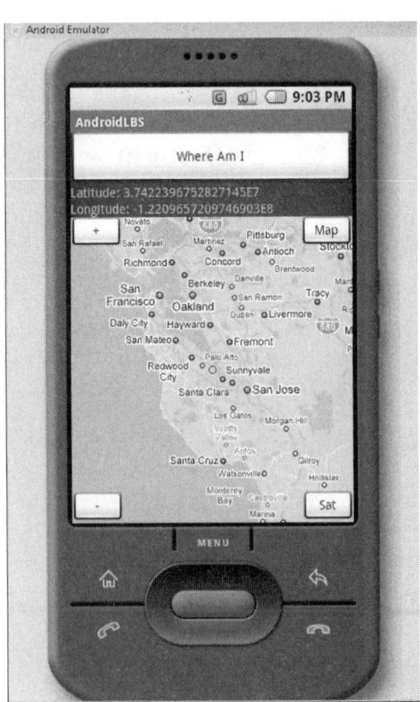

In the next chapter you will dive deeper into the Google API. Chapter 10 will walk you step by step through the process of using the Google API to send GTalk messages to and from an Android phone.

Ask the Expert

Q: Will the final release of Android continue to utilize .kml or .nmea files?

A: While the final release of Android was not finished at the time this book was written, it can be assumed that, yes, the final release will continue to utilize .kml and/or .nmea files. This would allow application developers to include files containing static coordinates with their applications.

Q: Is it possible to create a Google Map that has markers on it?

A: Yes, in Chapter 11 you will learn how to manipulate Google Map Overlays. These Views allow you to draw text, markers, and other shapes on top of Google Maps.

Chapter 10

Using the Google API with GTalk

Key Skills & Concepts

- Implementing a Google API package
- Configuring the XMPP development settings for Google access
- Implementing the View.OnClickListener( ) method

Chapter 9 introduced you to the Google API. You created an Activity that leveraged the Google API and Google Maps. Because of the ease and flexibility of the API, you were able to quickly display a Google Map of a user's current location. You also learned how to manipulate that map with relatively few lines of code.

The Google API contains more than just hooks into Google Maps. You used a small part of a much larger API in the last chapter. The base package for the Google API is com.google. From this base, the Google API contains packages that allow you to create Activities that leverage the power of GTalk (Google's chat service), Google Calendar, Google Docs, Google Spreadsheet, and Google Services.

When I started writing this book, the version of the Android SDK was m3-rc22. By the time I completed writing, Google had released m5-15. In the time between these two releases, Google had deprecated a few of these packages—while leaving them in the SDK.

Google Calendar, Google Spreadsheet, and Google Services appear to be undergoing an upgrade that, unfortunately, leaves them in a state of incompletion for the m5-rc15 release of the SDK. Google also removed any associated help files from the SDK for these packages, to avoid any confusion. Therefore, the focus of this chapter is a package that works quite well with the latest release of the Android SDK—GTalk.

In this chapter, you will build a small Activity that utilizes the GTalk package of the Android SDK. When the Activity is complete, you will be able to send GTalk messages from your phone to other GTalk users and receive messages from them.

NOTE

In the first iteration of the Google API for Android, the package dealing with GTalk was a much broader XMPP package. (XMPP is the protocol on which many chat platforms are based, including GTalk and Jabber.) With the latest release of the SDK, the original XMPP package was tightened up and renamed to reflect the specificity of GTalk.

To get started, create a new Project in Eclipse and name **GoogleAPI**.

Configuring the Android Emulator for GTalk

Before you can begin coding this project, you need to adjust a development setting on the Android Emulator, XMPP Settings.

With the project open, you need to depart from your routine for a minute. If you are familiar with GTalk, you are aware that you can use the product only when you log into your Google account. Therefore, you must take an extra step now to ensure that your device (in this case, the Android Emulator) can log into your Google account, thus enabling you to send and receive messages.

Navigate to your AndroidSDK/tools folder and launch the Emulator. You could launch it from within the Eclipse development environment, but that would require you to also launch an Activity that you have not coded yet. To save some time, just launch the Emulator manually.

After the Emulator is open, click the All shortcut. Find the Dev Tools item and launch it. You should see a menu similar to that shown in the following illustration.

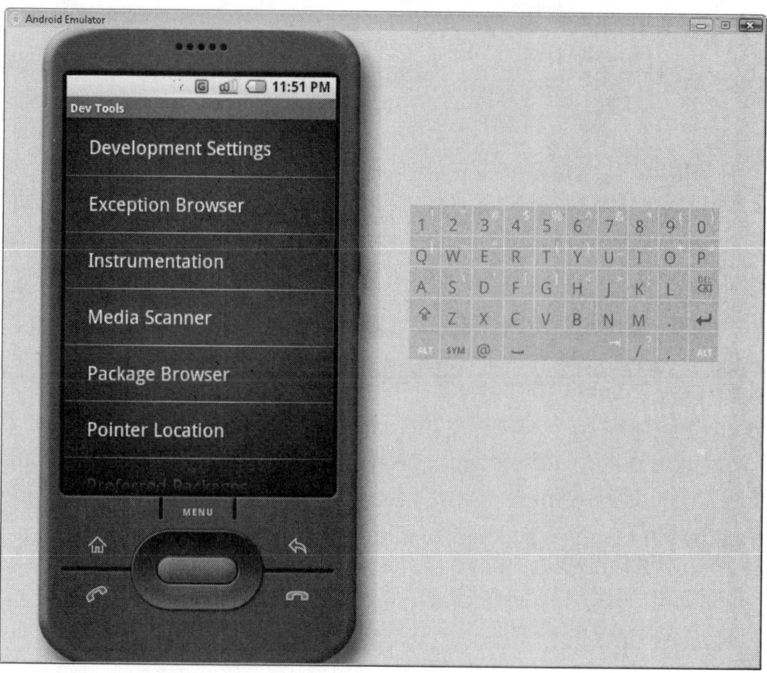

Scroll through the Dev Tools menu until you find XMPP Settings. Select XMPP Settings and you should see the Activity shown in the following illustration.

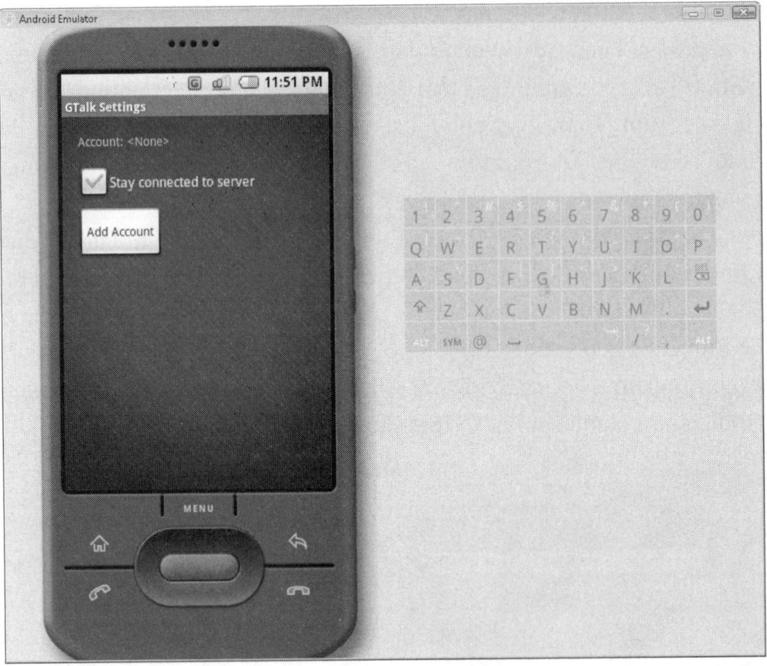

NOTE

When you open XMPP Settings, the Activity name is *GTalk Settings*. This may be an indication of where Google is going with the remaining packages of the Google API. The noticeable disconnect in the naming may be a leftover from the changes that were made between SDK versions.

The Activity should read Account:<None>, as the illustration shows. This indicates that there is not login information stored for your device. You need to add the login information for your Google account to allow your Activity access to Google's servers.

Click Add Account to display a screen, similar to the following, that you add your information to.

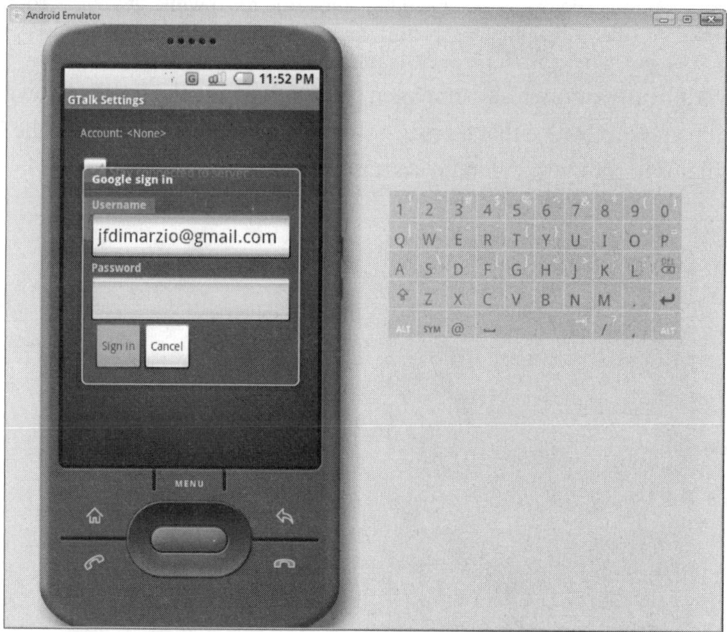

After you input your Username and Password, click Sign In. The Android Emulator should now attempt to authenticate your information. While the Emulator attempts to authenticate your information, it shows an "Authenticating" message.

CAUTION

Depending on your connection and whether or not you have a debugger connected to your Emulator, you may see this "Authenticating" message for a while. If your account is not authenticated after a few minutes, restart your Emulator and try again.

Once your information is authenticated, you should see the screen shown in the following illustration. Notice that there is no Return button here; just click the Home key on the Emulator to return to the main screen.

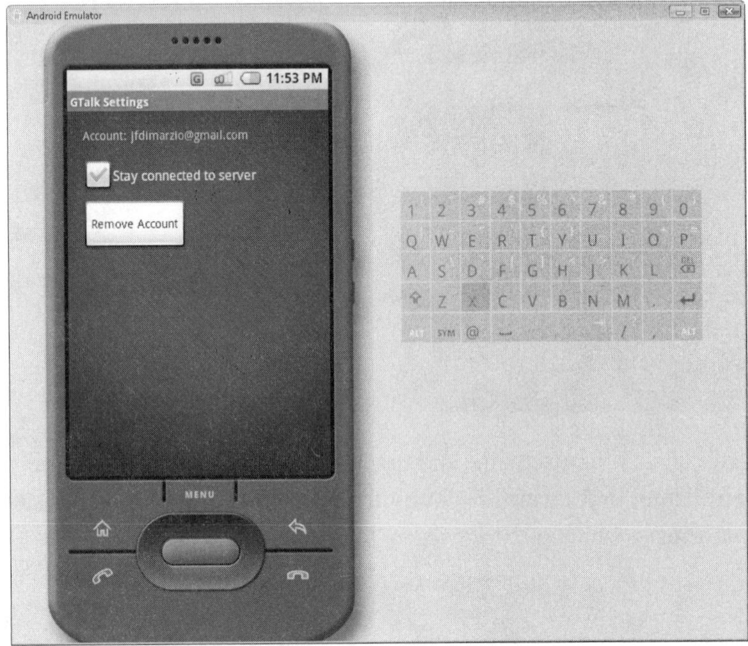

Now that the Emulator is configured and the project is set up, you can begin to code your Activity.

Implementing GTalk in Android

In this section you will use the Google API to create a GTalk-enabled Activity. This Activity will send and receive messages from the GTalk network, save them on screen, and display them in a notification bar. Your Activity will be able to communicate with other GTalk users, whether they are using GTalk on an Android phone or the PC.

The next section starts you off by creating the layout for the application. The first step in coding this new Activity is to add your layouts to GoogleAPI.xml.

Creating the Activity's Layout in the GoogleAPI.xml

This Activity consists of several Views. You need a ListView to display your text messages as you send and receive them. You also need two EditText Views, for the recipient's address and the message, and a Button for the send function.

First, set up a ListView with the id of messageList, as follows. You will be using a new attribute in this layout, *android:scrollbars*. Setting this attribute to *vertical* gives you a way to scroll through the message list.

```
<ListView
      android:id="@+id/messageList"
      android:layout_width="fill_parent"
      android:layout_height="0dip"
      android:scrollbars="vertical"
      android:layout_weight="1"
      android:drawSelectorOnTop="false" />
```

Place this ListView in the main layout tag. Under the ListView layout, place the layout for an EditText, as follows. This EditText will hold the address of the recipient that you are messaging.

```
<EditText
      android:id="@+id/messageTo"
      android:layout_width="wrap_content"
      android:layout_height="wrap_content"
      android:textSize="16sp"
      android:minWidth="250dp"
      android:scrollHorizontally="true" />
```

There should be nothing out of the ordinary with this EditText View.

Finally, you need to create a new horizontal layout to hold the message contents, EditText View, and the Send Button:

```
<LinearLayout
      xmlns:android="http://schemas.android.com/apk/res/android"
      android:orientation="horizontal"
      android:layout_width="fill_parent"
      android:layout_height="wrap_content">
```

```
<EditText
      android:id="@+id/messageText"
      android:layout_width="wrap_content"
      android:layout_height="wrap_content"
      android:textSize="16sp"
      android:minWidth="250dp"
      android:scrollHorizontally="true" />
<Button
      android:id="@+id/btnSend"
      android:layout_width="wrap_content"
      android:layout_height="wrap_content"
      android:text="Send Msg">
</Button>
</LinearLayout>
```

This layout will line up your Views so that they fall inline with each other. Place this LinearLayout in your main LinearLayout. Your final GoogleAPI.xml file should look like this:

```
<?xml version="1.0" encoding="utf-8"?>
<LinearLayout
      xmlns:android="http://schemas.android.com/apk/res/android"
      android:orientation="vertical"
      android:layout_width="fill_parent"
      android:layout_height="fill_parent">
<ListView
      android:id="@+id/messageList"
      android:layout_width="fill_parent"
      android:layout_height="0dip"
      android:scrollbars="vertical"
      android:layout_weight="1"
      android:drawSelectorOnTop="false" />
<EditText
      android:id="@+id/messageTo"
      android:layout_width="wrap_content"
      android:layout_height="wrap_content"
      android:textSize="16sp"
      android:minWidth="250dp"
      android:scrollHorizontally="true" />
<LinearLayout
      xmlns:android="http://schemas.android.com/apk/res/android"
      android:orientation="horizontal"
      android:layout_width="fill_parent"
```

```
        android:layout_height="wrap_content">
<EditText
        android:id="@+id/messageText"
        android:layout_width="wrap_content"
        android:layout_height="wrap_content"
        android:textSize="16sp"
        android:minWidth="250dp"
        android:scrollHorizontally="true" />
<Button
        android:id="@+id/btnSend"
        android:layout_width="wrap_content"
        android:layout_height="wrap_content"
        android:text="Send Msg">
</Button>
</LinearLayout>
</LinearLayout>
```

Adding Packages to GoogleAPI.java

With the layout file complete, there are a number of new packages you need to add to GoogleAPI.java. The first packages that you must import correspond to the Views you added to the layout. Therefore, you must import the packages for the EditText, ListView, ListAdapter, and Button:

```
import android.widget.EditText;
import android.widget.ListView;
import android.widget.ListAdapter;
import android.widget.Button;
```

You also need to import the packages of the Google API that deal with GTalk:

```
import com.google.android.gtalkservice.IGTalkSession;
import com.google.android.gtalkservice.IGTalkService;
import com.google.android.gtalkservice.GTalkServiceConstants;
import com.google.android.gtalkservice.IChatSession;
```

Some other packages that you need in this application include Intent, ServiceConnection, Color, and Im. The full list is as follows:

```
import android.content.ComponentName;
import android.content.Intent;
import android.content.ServiceConnection;
import android.database.Cursor;
```

```
import android.os.Bundle;
import android.os.DeadObjectException;
import android.os.IBinder;
import android.provider.Im;
import android.graphics.Color;
import android.view.View;
import android.widget.SimpleCursorAdapter;
```

As you can see, quite a few packages are needed for this Activity. However, as you will find, the amount of code needed to send and receive a message is relatively small. Now you need to implement an OnClickListener that will run your code.

Implementing the View.OnClickListener

You need to modify the GoogleAPI class to implement the View.OnClickListener. This will allow you to call the onClick( ) method from the Activity's main class when any button is clicked. Normally, this way of implementing the onClick( ) method is only effective when you have numerous buttons on one Activity and want to handle all the onClick calls in one method. However, I felt that you should still see how the method works so you can use it in your own future code. Keep in mind that I am showing this method because it can be a valuable tool in many situations.

```
public class GoogleAPI extends Activity implements View.OnClickListener {
}
```

Implementing general variables in your Activity is another concept that hasn't been covered previously in this book. You need to establish in this Activity a few general variables that you can work with from multiple methods:

```
    EditText messageText;
    ListView messageList;
    IGTalkSession myIGTalkSession;
    EditText messageTo;
    Button sendButton;
```

In your onCreate( ) method, you will perform your normal initializations. You should assign your layouts to your Views and set IGTalkSession to null. Also, just to add a little bit of interest to your Activity, change the background color of the ListView to gray.

```
        myIGTalkSession = null;
        messageText = (EditText) findViewById(R.id.messageText);
        messageList = (ListView) findViewById(R.id.messageList);
        messageTo = (EditText) findViewById(R.id.messageTo);
        sendButton = (Button) findViewById(R.id.btnSend);
        sendButton.setOnClickListener(this);
        messageList.setBackgroundColor(Color.GRAY );
```

TIP

Because you are implementing View.OnClickListener from your class, you can set the Send Button's OnClickListener( ) method to *this*.

The final piece of business to perform in the onCreate( ) method is to bind your service. This process creates the connection that you will use, facilitated by the Google account you established in the Dev Tools, to pass your GTalk messages:

```
this.bindService(new
Intent().setComponent(GTalkServiceConstants.GTALK_SERVICE_COMPONENT),
connection, 0);
```

In the bindService statement above, one of the parameters you pass to the setComponent( ) method is *connection*. This variable represents a ServiceConnection that implements the onServiceConnected( ) and onServiceDisconnected( ) methods. The following code builds the connection that is bound in the previous bindService statement:

```
     private ServiceConnection connection = new ServiceConnection() {
      public void onServiceConnected(ComponentName name, IBinder service) {
            try {
            myIGTalkSession =
IGTalkService.Stub.asInterface(service).getDefaultSession();
            } catch (DeadObjectException e) {
                  myIGTalkSession = null;
            }
            }
            public void onServiceDisconnected(ComponentName name) {
                  myIGTalkSession = null;
            }
    };
```

In the onServiceConnected( ) method, you establish a session using the IGTalkService.Stub. If this process fails, you should set the session to null once again. Similarly, in the onServiceDisconnected( ) method, you set the session to null.

Now you can create the code for the class's onClick event. There are several actions that you should perform during each onClick event:

1. Check the database for any messages.
2. Create a ListAdapter from the results of this query and display them to the ListView.
3. Create a ChatSession to the address in the EditView and send your message text.

NOTE

The Android server includes a SQLite database that you can use to hold many Activity-related items and any custom data you feel should be put into it. This database is introduced in depth in Chapter 11.

The following line of code queries the database for any messages sent between you and the messageTo recipient:

```
Cursor cursor = managedQuery(Im.Messages.CONTENT_URI, null,
"contact=\'" + messageTo.getText().toString() + "\'", null, null);
```

Use the following code to create a ListAdapter from the query results and assign the adapter to the ListView. You have used a similar process in a previous Activity, so it should not look foreign to you.

```
ListAdapter adapter = new SimpleCursorAdapter(this,
android.R.layout.simple_list_item_1, cursor,
new String[]{Im.MessagesColumns.BODY},
new int[]{android.R.id.text1});
this.messageList.setAdapter(adapter);
```

With the messages displayed, the last step is to send your message. The following lines of code create an IChatSession with the specified messageTo address. The message text is then passed over this session to the recipient.

```
try {
IChatSession chatSession;
chatSession =
myIGTalkSession.createChatSession(messageTo.getText().toString(););
```

```
chatSession.sendTextMessage(messageText.getText().toString());
            } catch (DeadObjectException ex) {
                  myIGTalkSession = null;
            }
```

When you put it all together, the complete GoogleAPI.java file should look like this:

```
package android_programmers_guide.GoogleAPI;

import android.app.Activity;
import android.content.ComponentName;
import android.content.Intent;
import android.content.ServiceConnection;
import android.database.Cursor;
import android.os.Bundle;
import android.os.DeadObjectException;
import android.os.IBinder;
import android.provider.Im;
import android.graphics.Color;
import android.view.View;
import android.widget.EditText;
import android.widget.ListView;
import android.widget.ListAdapter;
import android.widget.Button;
import android.widget.SimpleCursorAdapter;
import com.google.android.gtalkservice.IGTalkSession;
import com.google.android.gtalkservice.IGTalkService;
import com.google.android.gtalkservice.GTalkServiceConstants;
import com.google.android.gtalkservice.IChatSession;

public class GoogleAPI extends Activity implements View.OnClickListener {
       EditText messageText;
       ListView messageList;
       IGTalkSession myIGTalkSession;
       EditText messageTo;
       Button mSend;
    @Override
    public void onCreate(Bundle icicle) {
        super.onCreate(icicle);
        setContentView(R.layout.main);
      myIGTalkSession = null;
        messageText = (EditText) findViewById(R.id.messageText);
        messageList = (ListView) findViewById(R.id.messageList);
        messageTo = (EditText) findViewById(R.id.messageTo);
        mSend = (Button) findViewById(R.id.btnSend);
        mSend.setOnClickListener(this);
        messageList.setBackgroundColor(Color.GRAY );
```

```
        this.bindService(new
Intent().setComponent(GTalkServiceConstants.GTALK_SERVICE_COMPONENT),
connection, 0);

    }
    private ServiceConnection connection = new ServiceConnection() {
      public void onServiceConnected(ComponentName name, IBinder service) {
            try {
            myIGTalkSession =
IGTalkService.Stub.asInterface(service).getDefaultSession();
            } catch (DeadObjectException e) {
                  myIGTalkSession = null;
            }
            }

            public void onServiceDisconnected(ComponentName name) {
                  myIGTalkSession = null;
            }

    };
    public void onClick(View view) {

            Cursor cursor = managedQuery(Im.Messages.CONTENT_URI, null,
                    "contact=\'" + messageTo.getText().toString() + "\'",
null, null);
            ListAdapter adapter = new SimpleCursorAdapter(this,
                    android.R.layout.simple_list_item_1, cursor,
                    new String[]{Im.MessagesColumns.BODY},
                    new int[]{android.R.id.text1});
            this.messageList.setAdapter(adapter);

            try {
                  IChatSession chatSession;
                  chatSession =
myIGTalkSession.createChatSession(messageTo.getText().toString());
             chatSession.sendTextMessage(messageText.getText().toString());
            } catch (DeadObjectException ex) {
                  myIGTalkSession = null;
            }
        }

}
```

Compiling and Running GoogleAPI

Now, compile and run your GoogleAPI Activity in the Emulator. If your connection is successful, you should see a screen that looks like the following.

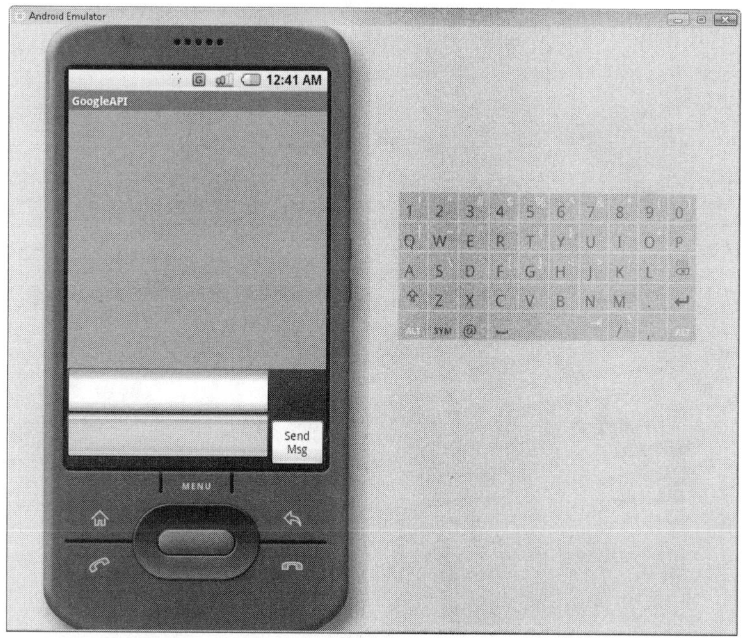

To test the Activity, I sent the message "Hello" to androidprogrammersguide@gmail.com, as shown here:

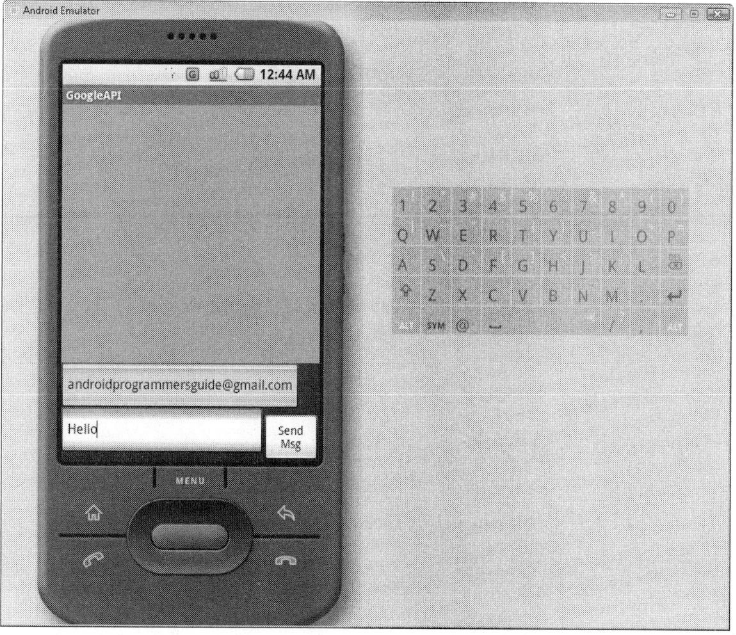

In the next illustration, you can see that clicking the Send Msg Button moves the message I sent to the ListView of messages.

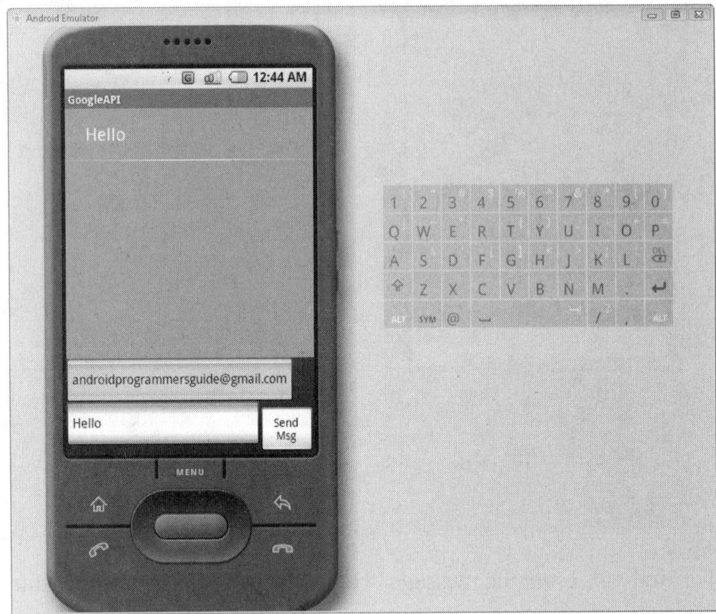

When I log in as androidprogrammersguide, I find that the message did indeed get passed through to the intended recipient, as shown here:

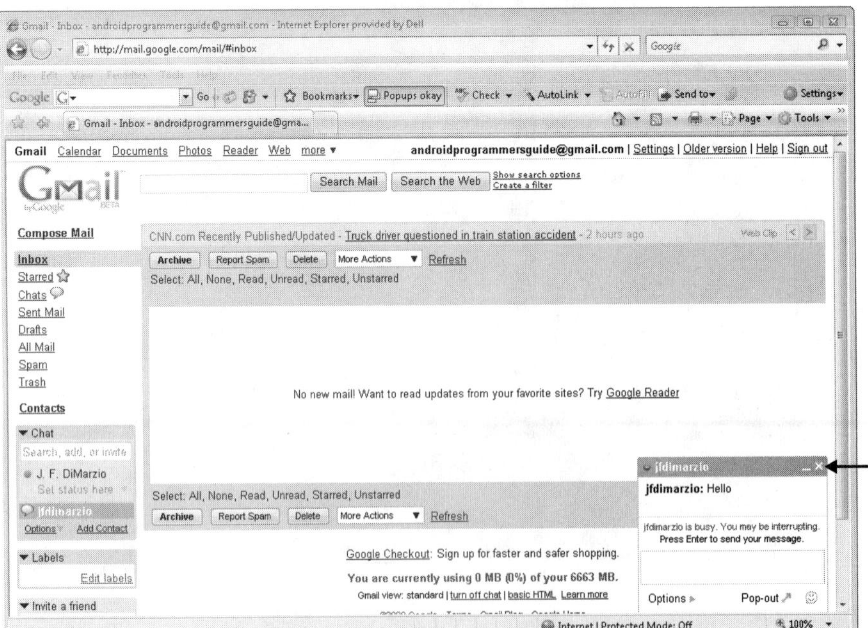

I replied to the chat with the text "Greetings!" To see this, look at the following two illustrations. Pay attention to the information bar at the top of the Activity screen. In the illustrations that follow, you can see that message is displayed with the sender.

In the next chapter, you will create your final application, in which you will use both the SQLite database and Google Maps Overlays to plot data records on a Google Map. These are very powerful technologies that elevate Android above other mobile operating systems.

Try This Add a Settings Feature to Your GoogleAPI Activity

Edit your GoogleAPI Activity to include a *settings* feature. Using the AndroidViews Activity from Chapter 8 as a guide, add a Button to the GoogleAPI Activity that can change the layout attributes of the application. Here are some ideas for what you may want a settings Button to do:

- Change the font of the message list

- Change the font color in the message list for messages you send as opposed to messages you receive
- Change the background color of the message list

Ask the Expert

Q: Can the GTalk API be used to communicate with other XMPP-based chat clients?

A: The answer to this is still unclear. The m3-rc22 version of the SDK included an XMPP API rather than the more specific GTalk API included in the m5-15 SDK. It is possible that these two will be combined in a future release of the Android SDK; in which case the GTalk API can be used to communicate with other XMPP-based chat clients.

Chapter 11

Application: Find a Friend

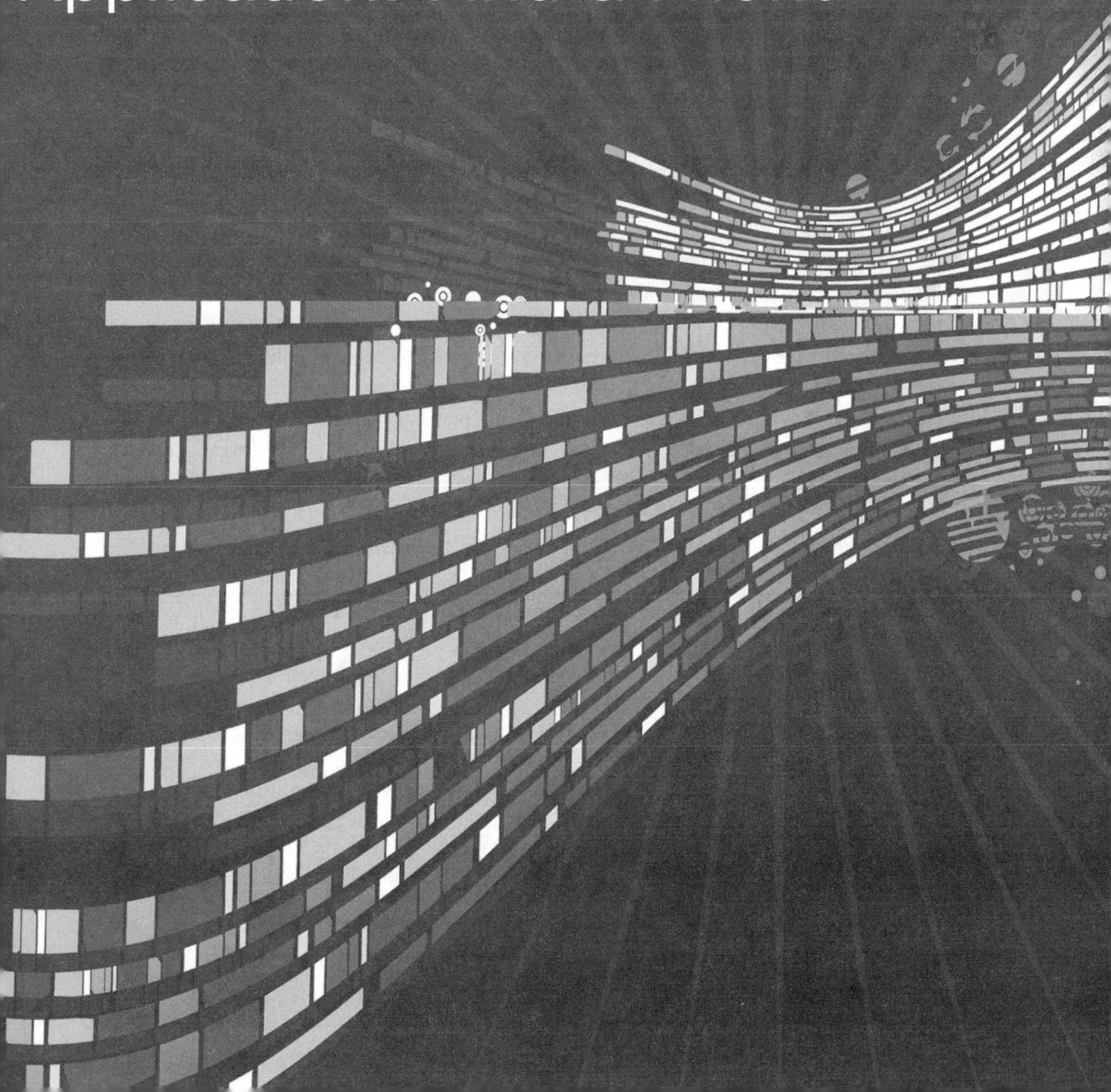

Key Skills & Concepts

- Create a SQLite database
- Create a custom Content Provider
- Retrieve items from a database and pass them to a Google Maps Overlay

This is the final chapter in which you will create an application, but it will be one of the largest applications introduced in this book. I will cover a couple of topics that you have not encountered thus far, and you will use the skills introduced in those topics to create a very robust application.

In this chapter, you will learn how to create SQLite databases on your Android Emulator. I will show you how to read, write, and delete data within your custom database. This process includes creating and using your own custom Content Provider to work with your database. Then, you will take the data that is stored in your database and write it out to a Google Maps Overlay. While you have worked with Google Maps previously in this book, you have not yet used an Overlay. You use Google Maps Overlays to draw shapes and write text on your map, resulting in very informative maps.

In this project, you will create a two-part application. The first part of the application will allow the user to enter "friends" into a mobile database. (A friend consists of a name and a geographic coordinate location.) The user will be able to add, modify, and delete friends.

The second part of the application will include a menu item. When the user selects this menu item, the application will display a Google Map. What make this different from the other Google Map you created in Chapter 9 is that this map will include a Google Maps Overlay, which enables you to write names, give information, and draw items on top of a Google Maps tile.

To begin, create a new Android Project within Eclipse named **FindAFriend**, using the settings shown in the following illustration.

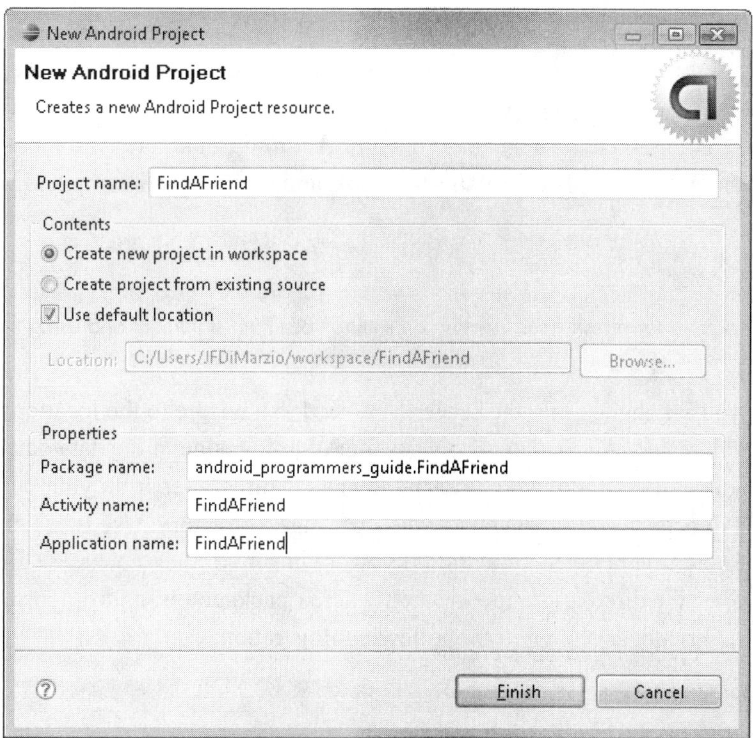

While you should be fairly comfortable creating Android applications by now, you will have a little bit of help creating this project. Google includes in the Android SDK an application called NotePad, a simple interface that lets you store, modify, and delete "notes" in a database. You are going to modify some of this sample code to create the interface for your Friends database.

If you want to see how Google NotePad works, load the project into Eclipse and run it in your Android Emulator before you move on. You will begin to modify this code shortly, but first, in the following section, you will create your first SQLite database.

Creating a SQLite Database

Android devices will ship with an internal SQLite database. The purpose of this database is to give users and developers a location in which to store information that can be used in Activities.

If you have used Microsoft SQL Server or SQLite, the structure and process for using Android's SQLite database will not seem foreign. Whatever your experience, this section covers all the skills you need to create and use a fully functional SQLite database.

You are going to create a database on your Android Emulator. To do this, you need to access the Android SDK command-line tools and use the **shell** command to access the Android server.

TIP

Refer to Chapter 3 to refresh your memory on setting your Path statement and using the command-line tools.

Once you are shelled into the server, you need to navigate to the location where the database will reside. All SQLite databases for Android reside in the data/data/<package>/ databases directory. Use the **cd** command to change directories from your current location to the data directory, and then again to the <package> directory. Use **ls** to list the files and directories at your current location if you are unsure of the <package> directory name. Change the directory to the location where <package> is, android_programmers_ guide.FindAFriend, as shown in the following illustration.

```
C:\Windows\system32\cmd.exe - adb shell
# cd data
cd data
# cd data
cd data
# ls
ls
android_programmers_guide.AndroidLBS
android_programmers_guide.AndroidViews
android.programmers.guide
android_programmers_guide.AndroidPhoneDialer
android_programmers_guide.GoogleAPI
android_programmers_guide.AndroidCamera
android
com.google.android.providers.contacts
com.google.android.providers.googleapps
com.google.android.providers.im
com.google.android.providers.media
com.google.android.providers.telephony
com.google.android.providers.settings
com.google.android.browser
com.google.android.contacts
com.google.android.development
com.google.android.googleapps
com.google.android.fallback
com.google.android.home
com.google.android.maps
com.google.android.phone
com.google.android.xmppService
com.google.android.xmppSettings
com.google.android.samples
com.google.android.masfproxyservice
com.google.android.gtalksettings
com.google.android.gtalkservice
android.screenshot
com.google.android.lunarlander
com.google.android.helloactivity
android_programmers_guide.GoogleApps
com.google.android.skeletonapp
android_programmers_guide.HelloWorldText
com.google.android.snake
com.google.android.notepad
android_programmers_guide.HelloWorldImage
android_programmers_guide.FindAFriend
#
```

CAUTION

If you do not have an android_programmers_guide.FindAFriend directory, create your application as explained in the previous section and quickly run the "Hello World!" default application that is created with your project. This will ensure that you have the correct directory.

Once you have navigated to the android_programmers_guide.FindAFriend directory, run the **ls** command. This command lists all the files and directories within a specific directory. This command should come up empty. As of right now, there are no files or directories inside your android_programmers_guide.FindAFriend directory.

Given that SQLite databases must be in a *databases* directory within this directory, this is a good time to create one. The tool **mkdir** creates directories for you. Therefore, run the command **mkdir databases**. This creates the directory that will hold your database.

CAUTION

Right now, you are most likely shelled into the server as root. Therefore, the directory you just created will be accessible only to root. This will prove very problematic when you attempt to run your Activity, because each Activity has a different user. To get around this, for development purposes, run **chmod 777 databases** to grant everyone access to the databases directory. However, in the future, you must be cautious about granting everyone rights to sensitive items on Android. Give to specific users only those rights that they need for specific items.

Now that you have created the directory for the database, you can create the database. Use the **cd** command to navigate into your databases directory. After you are in the databases directory, use the sqlite3 tool to create your database and name it friends.db, as follows:

```
# sqlite friends.db
```

If the command is successful, you should see a SQLite3 version message, in this case 3.5.0, and a SQLite3 prompt—sqlite>. This indicates that the database itself has been successfully created but is still empty. The database contains no tables or data. With this in mind, your next step is to create a table for your Activity's data.

You need to create a table called *friends*. This table will hold id, name, location, created, and modified fields. These fields will offer more than enough information for your project.

TIP

If you are not familiar with SQLite, a SQLite command must terminate with a semicolon. This is helpful if you want to span commands across prompts. Pressing the ENTER key without terminating a SQLite command will give you a continuation prompt, ...>. You can continue to enter your command at this prompt until you use the semicolon. SQLite will treat such continued commands as one full command once the semicolon is used.

To create your friends table within your database, enter the following command at the sqlite> prompt:

```
CREATE TABLE friends (_id INTEGER PRIMARY KEY, name TEXT, location TEXT,
created INTEGER, modified INTEGER);
```

If your command executes successfully, you will be returned to the sqlite> prompt, as shown in the following illustration.

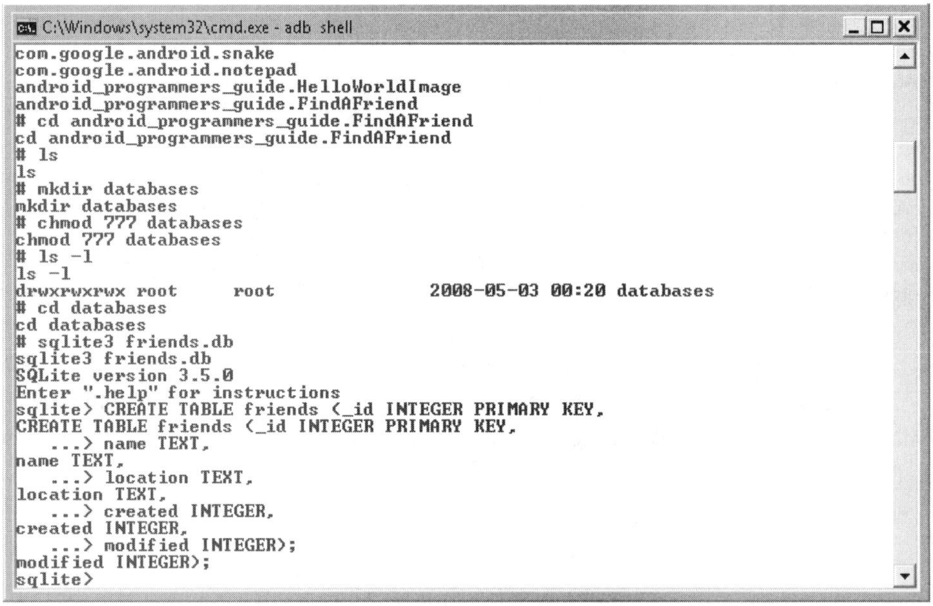

```
C:\Windows\system32\cmd.exe - adb shell
com.google.android.snake
com.google.android.notepad
android_programmers_guide.HelloWorldImage
android_programmers_guide.FindAFriend
# cd android_programmers_guide.FindAFriend
cd android_programmers_guide.FindAFriend
# ls
ls
# mkdir databases
mkdir databases
# chmod 777 databases
chmod 777 databases
# ls -l
ls -l
drwxrwxrwx root     root              2008-05-03 00:20 databases
# cd databases
cd databases
# sqlite3 friends.db
sqlite3 friends.db
SQLite version 3.5.0
Enter ".help" for instructions
sqlite> CREATE TABLE friends (_id INTEGER PRIMARY KEY,
CREATE TABLE friends (_id INTEGER PRIMARY KEY,
   ...> name TEXT,
name TEXT,
   ...> location TEXT,
location TEXT,
   ...> created INTEGER,
created INTEGER,
   ...> modified INTEGER);
modified INTEGER);
sqlite>
```

Your database is now ready to be used, and you can exit SQLite. Use the command **.exit** to exit. You can then quit your shell session and return to Eclipse.

Creating the database was the first step in setting up your application. Now that the database and corresponding table are created, you need a method to access the data. The data access method employed by Android is a Content Provider. The following section walks you through creating a custom Content Provider for your new database and accessing your data.

Creating a Custom Content Provider

Android uses Content Providers to mitigate access to data. You used a Content Provider in Chapter 9 to access and read coordinate information from a GPS. The same process applies to databases. There are Content Providers for Contact Lists, IMs, and Recent Calls. However, there is not yet a Content Provider for your Friends database.

Android is extremely flexible and allows you to create your own custom Content Providers for your own custom data. In this section you will create a Content Provider that works with your Friends database. This will be the key to access the friend data and eventually display it to the screen.

In the next section you will edit the strings.xml file. This file holds some global string content that can be used throughout your Activity.

Editing the strings.xml

First, you will edit the strings.xml file for your project. The strings.xml file is created with each project but you have not used it yet. This file holds static strings that can be used in your Activities.

Typically, you will not be able to determine all of the strings that you need to use in your Activity before you even write it. That is, you will usually add entries to strings.xml as you build the Activity. However, because that would break the flow of the book, I am giving you the full contents of the strings.xml file up front. Edit your strings.xml file to look as follows:

```
<?xml version="1.0" encoding="utf-8"?>
<resources>
    <string name="app_name">FindAFriend</string>
    <string name="menu_delete">Delete</string>
    <string name="menu_insert">Add Friend</string>
    <string name="find_friends">Find Friends</string>
    <string name="menu_revert">Revert</string>
    <string name="menu_discard">Discard</string>
    <string name="resolve_edit">Edit location</string>
    <string name="resolve_title">Edit name</string>
    <string name="title_create">Create Friend</string>
    <string name="title_edit">Edit Friend</string>
      <string name="title_notes_list">Friends</string>
      <string name="title_note">Location</string>
      <string name="title_edit_title">Friend Name:</string>
      <string name="button_ok">OK</string>
      <string name="error_title">Error</string>
```

```
    <string name="error_message">Error loading note</string>
</resources>
```

With the strings.xml file complete, you need to create a .java file to hold your code. You should call this file **FriendsProvider.java**. You also need to create another .java file to hold your data definition. Name this file **Friends.java**, because it will define what a *Friends* data structure looks like and let your Content Provider access it correctly. (Because the provider will be a class that sits in your project, there is no need to build a corresponding .xml layout file.)

TIP

Technically, your custom Content Provider does not need to reside within the same project or package as the rest of the code for this application. For simplicity's sake, I am making the Content Provider a class in the FindAFriend project. However, if you plan to create a Content Provider that may be used by multiple projects, create it in a separate package. This will let you call one package when you want to use the Content Provider only.

Let's start with the Friends.java file. You need to import only two packages for this relatively small class:

```
import android.net.Uri;
import android.provider.BaseColumns;
```

BaseColumns will be implemented by a subclass off of your main Friends class. Name this subclass **Friend**, because it will represent one friend from the Friends dataset. The following code shows how you should set up the class outline:

```
public final class Friends {
    public static final class Friend implements BaseColumns {
}
}
```

This class will hold some static variables that define each of the columns in your Friends database, the Content URI, and the default sort order for the records.

TIP

A Content URI is used to identify the content that you will handle. This value must be unique.

The strings that you need to define look like this:

```
        public static final Uri CONTENT_URI
                =
Uri.parse("content://android_programmers_guide.FindAFriend.Friends/friend");

        public static final String DEFAULT_SORT_ORDER = "modified DESC";

        public static final String NAME = "name";

        public static final String LOCATION = "location";

        public static final String CREATED_DATE = "created";

        public static final String MODIFIED_DATE = "modified";
```

With these variables set, the contents of your Friends class come together perfectly. The full file should look like this:

```
package android_programmers_guide.FindAFriend;

import android.net.Uri;
import android.provider.BaseColumns;

public final class Friends {
    public static final class Friend implements BaseColumns {
        public static final Uri CONTENT_URI
                =
Uri.parse("content://android_programmers_guide.FindAFriend.Friends/friend");

        public static final String DEFAULT_SORT_ORDER = "modified DESC";

        public static final String NAME = "name";

        public static final String LOCATION = "location";

        public static final String CREATED_DATE = "created";

        public static final String MODIFIED_DATE = "modified";
    }
}
```

In the next section you will create your Content Provider.

Creating Your Content Provider

Using Eclipse, open FriendsProvider.java, which will become the Content Provider for your project. You are going to use this custom Content Provider in your Activity to retrieve data from your Friends database.

As always, let's start by looking at the imports for this file. You need to import the Friends class and several other classes:

```
import android_programmers_guide.FindAFriend.Friends;
import android.content.*;
import android.database.Cursor;
import android.database.SQLException;
import android.database.sqlite.SQLiteOpenHelper;
import android.database.sqlite.SQLiteDatabase;
import android.database.sqlite.SQLiteQueryBuilder;
import android.net.Uri;
import android.text.TextUtils;
import android.util.Log;
import java.util.HashMap;
```

As you can see, you are importing several packages here, most of which deal with SQL. I will explain these packages as you use them.

The package you will be using first is android.content. To utilize and override the required methods for being a Content Provider, your FriendsProvider class needs to extend ContentProvider. Take a look at the following class outline, which includes several variable definitions that you will use throughout your provider:

```
public class FriendsProvider extends ContentProvider {
    private SQLiteDatabase mDB;
    private static final String TAG = "FriendsProvider";
    private static final String DATABASE_NAME = "friends";
    private static final int DATABASE_VERSION = 2;

    private static HashMap<String, String> FRIENDS_PROJECTION_MAP;

    private static final int FRIENDS = 1;
    private static final int FRIENDS_ID = 2;
private static final UriMatcher URL_MATCHER;}
```

The Content Provider contains several methods that you will want to override, including onCreate( ), query( ), insert( ), delete( ), and update( ). Because these methods will be called by Activities using your Content Provider, you must override them to specifically access the Friends database.

The onCreate( ) method that you will be overriding calls a SQLiteOpenHelper. Therefore, before you can override the onCreate( ) method of the ContentProvider, you have to create a class that extends SQLiteOpenHelper.

The code block that follows is a subclass of your Content Provider that extends SQLiteOpenHelper:

```
private static class DatabaseHelper extends SQLiteOpenHelper {

@Override
 public void onCreate(SQLiteDatabase db) {
  db.execSQL("CREATE TABLE friends (_id INTEGER PRIMARY KEY,"
                  + "name TEXT," + "location TEXT," + "created INTEGER,"
                  + "modified INTEGER" + ");");
        }

@Override
  public void onUpgrade(SQLiteDatabase db, int oldVersion, int
newVersion) {
       Log.w(TAG, "Upgrading database from version " + oldVersion + "to "
        + newVersion + ", which will destroy all old data");
            db.execSQL("DROP TABLE IF EXISTS friends");
            onCreate(db);
    }
 }
```

The DatabaseHelper class you just created contains two overridden methods: onCreate( ) and onUpgrade( ). The onCreate( ) method is used when creating the database from code, or in instances where the table definition does not exist.

NOTE

Given that you created the database structure from the adb shell, you will not rely on the onCreate( ) method of DatabaseHelper to establish your database.

With the DatabaseHelper class created, you can now override the onCreate( ) method for your Content Provider:

```
    @Override
    public boolean onCreate() {
        DatabaseHelper dbHelper = new DatabaseHelper();
        mDB = dbHelper.openDatabase(getContext(), DATABASE_NAME, null,
DATABASE_VERSION);
        return (mDB == null) ? false : true;

    }
```

This is a fairly simple method that, in the end, returns a Boolean representing whether or not your database could be opened. You use the SQLiteOpenHelper created in your sibling class to open the Friends database. Notice that you pass the database name into the DatabaseHelper class. If the database object—mDB—is not null when it returns, then the database was successfully opened and you can query it.

Next, override the query( ) method of the ContentProvider class. This will be the meat of your Content Provider. The query( ) method is called from your Activity through the Content Provider to gather the records from your database. Take a look at the code in the overridden version of the query( ) method:

```
@Override
public Cursor query(Uri url, String[] projection, String selection,
        String[] selectionArgs, String sort) {
    SQLiteQueryBuilder qb = new SQLiteQueryBuilder();

    switch (URL_MATCHER.match(url)) {
    case FRIENDS:
        qb.setTables("friends");
        qb.setProjectionMap(FRIENDS_PROJECTION_MAP);
        break;

    case FRIENDS_ID:
        qb.setTables("friends");
        qb.appendWhere("_id=" + url.getPathSegments().get(1));
        break;

    default:
        throw new IllegalArgumentException("Unknown URL " + url);
    }

    String orderBy;
    if (TextUtils.isEmpty(sort)) {
        orderBy = Friends.Friend.DEFAULT_SORT_ORDER;
    } else {
        orderBy = sort;
    }

    Cursor c = qb.query(mDB, projection, selection, selectionArgs, null,
null, orderBy);
    c.setNotificationUri(getContext().getContentResolver(), url);
    return c;
}
```

The query( ) method does a little bit of housekeeping, by checking the validity of the database URL passed into it and defining a query sort order. The URL check is to ensure that you are trying to access only the Friends database. If you are attempting to access a database from another Activity, or from another Content Provider, the query( ) method throws an exception.

Toward the end of the method, you perform a query using a SQLiteQueryBuilder. The resulting dataset is assigned to a Cursor using the following line of code:

```
        Cursor c = qb.query(mDB, projection, selection, selectionArgs, null,
null, orderBy);
```

NOTE

A Cursor is a device that allows you to move through records and return information from columns.

The update( ), delete( ), and insert( ) methods are also fairly straightforward in design. Take a look at these three methods, as you should override them:

```
    @Override
    public Uri insert(Uri url, ContentValues initialValues) {
        long rowID;
        ContentValues values;
        if (initialValues != null) {
            values = new ContentValues(initialValues);
        } else {
            values = new ContentValues();
        }

        if (URL_MATCHER.match(url) != FRIENDS) {
            throw new IllegalArgumentException("Unknown URL " + url);
        }

        Long now = Long.valueOf(System.currentTimeMillis());
        Resources r = Resources.getSystem();

        if (values.containsKey(Friends.Friend.CREATED_DATE ) == false) {
            values.put(Friends.Friend.CREATED_DATE, now);
        }

        if (values.containsKey(Friends.Friend.MODIFIED_DATE) == false) {
            values.put(Friends.Friend.MODIFIED_DATE, now);
        }

        if (values.containsKey(Friends.Friend.NAME) == false) {
            values.put(Friends.Friend.NAME,
r.getString(android.R.string.untitled));
        }

        if (values.containsKey(Friends.Friend.LOCATION) == false) {
            values.put(Friends.Friend.LOCATION , "");
        }
```

```
        rowID = mDB.insert("friends", "friend", values);
        if (rowID > 0) {
            Uri uri = ContentUris.withAppendedId(Friends.Friend.CONTENT_URI
, rowID);
            getContext().getContentResolver().notifyChange(uri, null);
            return uri;
        }

        throw new SQLException("Failed to insert row into " + url);
    }

    @Override
    public int delete(Uri url, String where, String[] whereArgs) {
        int count;
        long rowId = 0;
        switch (URL_MATCHER.match(url)) {
        case FRIENDS:
            count = mDB.delete("friends", where, whereArgs);
            break;

        case FRIENDS_ID:
            String segment = url.getPathSegments().get(1);
            rowId = Long.parseLong(segment);
            count = mDB
                    .delete("friends", "_id="
                            + segment
                            + (!TextUtils.isEmpty(where) ? " AND (" + where
                                    + ')' : ""), whereArgs);
            break;

        default:
            throw new IllegalArgumentException("Unknown URL " + url);
        }

        getContext().getContentResolver().notifyChange(url, null);
        return count;
    }

    @Override
    public int update(Uri url, ContentValues values, String where, String[]
whereArgs) {
        int count;
        switch (URL_MATCHER.match(url)) {
        case FRIENDS:
            count = mDB.update("friends", values, where, whereArgs);
            break;

        case FRIENDS_ID:
            String segment = url.getPathSegments().get(1);
            count = mDB
                    .update("friends", values, "_id=" + segment
```

```
+ (!TextUtils.isEmpty(where) ? " AND (" + where + ')' : ""), whereArgs);
            break;

        default:
            throw new IllegalArgumentException("Unknown URL " + url);
        }

        getContext().getContentResolver().notifyChange(url, null);
        return count;
    }
```

The code within these methods should be fairly self-explanatory. If you look past the housekeeping that takes place in each method, the core of the code issues a database statement to perform the requested action of updating, deleting, or inserting.

The final part of the Content Provider will be a getType( ) method that returns the type of your Friends data. When creating your own type, you should always follow this convention:

```
vnd.android.cursor.dir/vnd.<package>
```

Take a look at the getType( ) method:

```
    @Override
    public String getType(Uri url) {
        switch (URL_MATCHER.match(url)) {
        case FRIENDS:
            return
"vnd.android.cursor.dir/vnd.android_programmers_guide.friend";

        case FRIENDS_ID:
            return
"vnd.android.cursor.item/vnd.android_programmers_guide.friend";

        default:
            throw new IllegalArgumentException("Unknown URL " + url);
        }
    }
```

That should complete your new custom Content Provider. Take a look at the completed FriendsProvider code:

```
package android_programmers_guide.FindAFriend;
import android_programmers_guide.FindAFriend.Friends;
import android.content.*;
```

```
import android.database.Cursor;
import android.database.SQLException;
import android.database.sqlite.SQLiteOpenHelper;
import android.database.sqlite.SQLiteDatabase;
import android.database.sqlite.SQLiteQueryBuilder;
import android.net.Uri;
import android.text.TextUtils;
import android.util.Log;
import java.util.HashMap;

public class FriendsProvider extends ContentProvider {
    private SQLiteDatabase mDB;
    private static final String TAG = "FriendsProvider";
    private static final String DATABASE_NAME = "friends";
    private static final int DATABASE_VERSION = 2;

    private static HashMap<String, String> FRIENDS_PROJECTION_MAP;

    private static final int FRIENDS = 1;
    private static final int FRIENDS_ID = 2;

    private static final UriMatcher URL_MATCHER;

    private static class DatabaseHelper extends SQLiteOpenHelper {

        @Override
        public void onCreate(SQLiteDatabase db) {
            db.execSQL("CREATE TABLE friends (_id INTEGER PRIMARY KEY,"
                    + "name TEXT," + "location TEXT," + "created INTEGER,"
                    + "modified INTEGER" + ");");
        }

        @Override
        public void onUpgrade(SQLiteDatabase db, int oldVersion, int
newVersion) {
       Log.w(TAG, "Upgrading database from version " + oldVersion + "to "
        + newVersion + ", which will destroy all old data");
            db.execSQL("DROP TABLE IF EXISTS friends");
            onCreate(db);
        }
    }

    @Override
    public boolean onCreate() {
        DatabaseHelper dbHelper = new DatabaseHelper();
        mDB = dbHelper.openDatabase(getContext(), DATABASE_NAME, null,
DATABASE_VERSION);
        return (mDB == null) ? false : true;

    }
```

```
    @Override
    public Cursor query(Uri url, String[] projection, String selection,
            String[] selectionArgs, String sort) {
        SQLiteQueryBuilder qb = new SQLiteQueryBuilder();

        switch (URL_MATCHER.match(url)) {
        case FRIENDS:
            qb.setTables("friends");
            qb.setProjectionMap(FRIENDS_PROJECTION_MAP);
            break;

        case FRIENDS_ID:
            qb.setTables("friends");
            qb.appendWhere("_id=" + url.getPathSegments().get(1));
            break;

        default:
            throw new IllegalArgumentException("Unknown URL " + url);
        }

        String orderBy;
        if (TextUtils.isEmpty(sort)) {
            orderBy = Friends.Friend.DEFAULT_SORT_ORDER;
        } else {
            orderBy = sort;
        }

        Cursor c = qb.query(mDB, projection, selection, selectionArgs, null,
null, orderBy);
        c.setNotificationUri(getContext().getContentResolver(), url);
        return c;
    }

    @Override
    public String getType(Uri url) {
        switch (URL_MATCHER.match(url)) {
        case FRIENDS:
            return
"vnd.android.cursor.dir/vnd.android_programmers_guide.friend";

        case FRIENDS_ID:
            return
"vnd.android.cursor.item/vnd.android_programmers_guide.friend";

        default:
            throw new IllegalArgumentException("Unknown URL " + url);
        }
    }

    @Override
```

```
    public Uri insert(Uri url, ContentValues initialValues) {
        long rowID;
        ContentValues values;
        if (initialValues != null) {
            values = new ContentValues(initialValues);
        } else {
            values = new ContentValues();
        }

        if (URL_MATCHER.match(url) != FRIENDS) {
            throw new IllegalArgumentException("Unknown URL " + url);
        }

        Long now = Long.valueOf(System.currentTimeMillis());
        Resources r = Resources.getSystem();

        if (values.containsKey(Friends.Friend.CREATED_DATE ) == false) {
            values.put(Friends.Friend.CREATED_DATE, now);
        }

        if (values.containsKey(Friends.Friend.MODIFIED_DATE) == false) {
            values.put(Friends.Friend.MODIFIED_DATE, now);
        }

        if (values.containsKey(Friends.Friend.NAME) == false) {
            values.put(Friends.Friend.NAME,
r.getString(android.R.string.untitled));
        }

        if (values.containsKey(Friends.Friend.LOCATION) == false) {
            values.put(Friends.Friend.LOCATION , "");
        }

        rowID = mDB.insert("friends", "friend", values);
        if (rowID > 0) {
            Uri uri = ContentUris.withAppendedId(Friends.Friend.CONTENT_URI
, rowID);
            getContext().getContentResolver().notifyChange(uri, null);
            return uri;
        }

        throw new SQLException("Failed to insert row into " + url);
    }

    @Override
    public int delete(Uri url, String where, String[] whereArgs) {
        int count;
        long rowId = 0;
        switch (URL_MATCHER.match(url)) {
        case FRIENDS:
```

```
            count = mDB.delete("friends", where, whereArgs);
            break;

        case FRIENDS_ID:
            String segment = url.getPathSegments().get(1);
            rowId = Long.parseLong(segment);
            count = mDB
                    .delete("friends", "_id="
                            + segment
                            + (!TextUtils.isEmpty(where) ? " AND (" + where
                                    + ')' : ""), whereArgs);
            break;

        default:
            throw new IllegalArgumentException("Unknown URL " + url);
        }

        getContext().getContentResolver().notifyChange(url, null);

        return count;
    }

    @Override
    public int update(Uri url, ContentValues values, String where, String[]
whereArgs) {
        int count;
        switch (URL_MATCHER.match(url)) {
        case FRIENDS:
            count = mDB.update("friends", values, where, whereArgs);
            break;

        case FRIENDS_ID:
            String segment = url.getPathSegments().get(1);
            count = mDB
                    .update("friends", values, "_id=" + segment
+ (!TextUtils.isEmpty(where) ? " AND (" + where + ')' : ""), whereArgs);
            break;

        default:
            throw new IllegalArgumentException("Unknown URL " + url);
        }

        getContext().getContentResolver().notifyChange(url, null);
        return count;
    }

    static {
        URL_MATCHER = new UriMatcher(UriMatcher.NO_MATCH);
        URL_MATCHER.addURI("android_programmers_guide.FindAFriend.Friends",
"friend", FRIENDS);
```

```
        URL_MATCHER.addURI("android_programmers_guide.FindAFriend.Friends",
"friend/#", FRIENDS_ID);

        FRIENDS_PROJECTION_MAP = new HashMap<String, String>();
        FRIENDS_PROJECTION_MAP.put(Friends.Friend._ID, "_id");
        FRIENDS_PROJECTION_MAP.put(Friends.Friend.NAME, "name");
        FRIENDS_PROJECTION_MAP.put(Friends.Friend.LOCATION, "location");
        FRIENDS_PROJECTION_MAP.put(Friends.Friend.CREATED_DATE, "created");
        FRIENDS_PROJECTION_MAP.put(Friends.Friend.MODIFIED_DATE,
"modified");
    }
}
```

With the underlying data elements now created (the database, definitions, and Content Provider), you can begin to build the surrounding Activity. Remember, this activity will use the data in your database, display it to a list, and then allow the user to launch another Activity that places database items on a Google Maps Overlay. In the following section, you will build both Activities and complete your FindAFriend application.

Creating the FindAFriend Activity

If you have taken the time to run Google's NotePad demo, then you will be very familiar with the layout of this Activity. You will be modifying the NotePad interface to work with your Friends database and Google Maps. The FindAFriend Activity will interact with several smaller Activities: NameEditor, LocationEditor, and FriendsMap. You will build all of these Activities in the coming sections.

NOTE

In addition to NotePad, Google provides several very well-written demo Activities that outline basic techniques for multiple programming situations.

As you have done with past Activity projects from this book, start with the AndroidManifest.xml file. Being a fairly complex application, you need to make multiple changes to AndroidManifest.xml.

Editing AndroidManifest.xml

Take a look at the following AndroidManifest.xml file for the FindAFriend project. You need to add several Intent Filters for new Activities, including ones to edit a friend's location, edit a friend's name, and launch your Google Map.

Also, pay close attention to the *actions* within each Intent Filter. These represent actions that will be passed to each Activity handling that Intent. Finally, do not forget to add Access_Location and Access_GPS permission so that you can add your current location to the map as well.

The full AndroidManifest.xml file should appear as follows:

```
<?xml version="1.0" encoding="utf-8"?>
<manifest xmlns:android="http://schemas.android.com/apk/res/android"
    package="android_programmers_guide.FindAFriend">
    <application android:icon="@drawable/icon">
            <provider android:name="FriendsProvider"

android:authorities="android_programmers_guide.FindAFriend.Friends" />
        <activity android:name=".FindAFriend"
android:label="@string/app_name">
            <intent-filter>
                <action android:name="android.intent.action.MAIN" />
                <category android:name="android.intent.category.LAUNCHER" />
            </intent-filter>
        <intent-filter>
                <action android:name="android.intent.action.VIEW" />
                <action android:name="android.intent.action.EDIT" />
                <action android:name="android.intent.action.PICK" />
                <category android:name="android.intent.category.DEFAULT" />
                <dataandroid:mimeType="vnd.android.cursor.dir/
vnd.android_programmers_guide.friend" />
            </intent-filter>
            <intent-filter>
                <action android:name="android.intent.action.GET_CONTENT" />
                <category android:name="android.intent.category.DEFAULT" />
                <dataandroid:mimeType="vnd.android.cursor.item/
vnd.android_programmers_guide.friend" />
            </intent-filter>
        </activity>
        <activity android:name=".FriendsMap" android:label="FriendsMap">
            <intent-filter>
                 <action android:name="android.intent.action.MAIN" />
                 <category android:name="android.intent.category.LAUNCHER"
/>
            </intent-filter>
        </activity>
        <activity android:name="LocationEditor"
android:label="@string/title_note">
            <intent-filter android:label="@string/resolve_edit">
                <action android:name="android.intent.action.VIEW" />
                <action android:name="android.intent.action.EDIT" />
                <action
android:name="com.google.android.notepad.action.EDIT_LOCATION" />
                <category android:name="android.intent.category.DEFAULT" />
                <dataandroid:mimeType="vnd.android.cursor.item/
vnd.android_programmers_guide.friend" />
            </intent-filter>
```

```
            <intent-filter>
                <action android:name="android.intent.action.INSERT" />
                <category android:name="android.intent.category.DEFAULT" />
                <dataandroid:mimeType="vnd.android.cursor.dir/
vnd.android_programmers_guide.friend" />
            </intent-filter>

        </activity>

            <activity android:name="NameEditor"
android:label="@string/title_edit_title"
                        android:theme="@android:style/Theme.Dialog">
            <intent-filter android:label="@string/resolve_title">
                <action android:name="com.google.android.notepad.action.EDIT_NAME"
/>
                <category android:name="android.intent.category.DEFAULT" />
                <category android:name="android.intent.category.ALTERNATIVE" />
<category android:name="android.intent.category.SELECTED_ALTERNATIVE" />
                <dataandroid:mimeType="vnd.android.cursor.item/
vnd.android_programmers_guide.friend" />
            </intent-filter>
        </activity>
    </application>
<uses-permission android:name="android.permission.ACCESS_GPS">
</uses-permission><uses-permission
android:name="android.permission.ACCESS_LOCATION">
</uses-permission></manifest>
```

In the next section, you will create the first Activity for this project—NameEditor. As the name implies, this Activity will be launched when the user wishes to edit the name of a friend.

Creating the NameEditor Activity

In this section, you will create the NameEditor Activity for the FindAFriend project. This Activity will be launched from a menu item on the main FindAFriend Activity (which you have not created yet). The purpose of the NameEditor Activity will be to modify the name field of a Friend record.

Add a name_editor.xml layout file and a corresponding NameEditor.java file to your application. You will edit these files to create your Activity.

First, edit name_editor.xml to create the layout for the Activity. The Activity will hold one EditText and one Button. The EditText will allow you to modify the name field, and the Button will write the results and exit. If you followed this book from the beginning, you have added quite a few View layouts to XML files. Therefore, I can spare you the details of each addition individually. The full name_editor.xml file should appear as follows:

```
<LinearLayout xmlns:android="http://schemas.android.com/apk/res/android"
      android:layout_width="wrap_content"
    android:layout_height="wrap_content"
    android:orientation="vertical"
    android:paddingLeft="6dip"
    android:paddingRight="6dip"
    android:paddingBottom="3dip">

    <EditText android:id="@+id/name"
        android:maxLines="1"
        android:layout_marginTop="2dip"
        android:layout_width="wrap_content"
            android:ems="25"
        android:layout_height="wrap_content"
        android:autoText="true"
        android:capitalize="sentences"
        android:scrollHorizontally="true" />

    <Button android:id="@+id/ok"
        android:layout_width="wrap_content"
        android:layout_height="wrap_content"
        android:layout_gravity="right"
        android:text="@string/button_ok" />

</LinearLayout>
```

Now, edit NameEditor.java and begin building your code. You need to import your Friends class from the previous sections and import the Cursor package to help you work with the database records:

```
import android.app.Activity;
import android.database.Cursor;
import android.net.Uri;
import android.os.Bundle;
import android.view.View;
import android.widget.Button;
import android.widget.EditText;
```

You should establish your Activity so that you implement the View.OnClickListener( ). This will let you override the OnClickListener( ) methods in your Activity. This code sample shows the outline of your NameEditor class with some variable definitions that you will need:

```
public class NameEditor extends Activity implements View.OnClickListener {

    public static final String EDIT_NAME_ACTION =
        "android_programmers_guide.FindAFriend.action.EDIT_NAME";
```

```
    private static final int NAME_INDEX = 1;

    private static final String[] PROJECTION = new String[] {
            Friends.Friend._ID,
            Friends.Friend.NAME,
    };

    Cursor mCursor;
    EditText mText;
}
```

Next, you need to override some methods, starting with onCreate( ). You have seen this method overridden in other chapters. Typically, it holds all the code that should be executed when the Activity is created.

```
public void onCreate(Bundle icicle) {
    super.onCreate(icicle);

    setContentView(R.layout.name_editor);

    Uri uri = getIntent().getData();

    mCursor = managedQuery(uri, PROJECTION, null, null);

    mText = (EditText) this.findViewById(R.id.name);
    mText.setOnClickListener(this);

    Button b = (Button) findViewById(R.id.ok);
    b.setOnClickListener(this);
}
```

Notice that, in the previous code sample, you assign layouts to their respective Views and initiate some of your variables. However, you may be wondering where the data is for the name field. That is, you have created a cursor, but you have not retrieved anything from it. You will use the onResume( ) method for that.

The two methods that you will override next, onResume( ) and onPause( ), will do the work of reading from and writing to the database, respectively. Within the Android life cycle, onResume( ) is called when an Activity is open and on the top of the focus. onPause( ) is called when an Activity is being closed but before focus is handed to another Activity.

Override your onResume( ) method to read the database and retrieve the name field:

```
protected void onResume() {
    super.onResume();
```

```
        if (mCursor != null) {
            mCursor.first();
            String title = mCursor.getString(NAME_INDEX);
            mText.setText(title);
        }
}
```

In this method, you move the Cursor to its first record, read the name field from it using the index assigned earlier, and set the EditText to the contents of the name field. This automatically populates the field with the current record's name value.

Next, modify the onPause( ) method to write the contents of the EditText back to the database:

```
    protected void onPause() {
        super.onPause();

        if (mCursor != null) {
            String title = mText.getText().toString();
            mCursor.updateString(NAME_INDEX, title);
            mCursor.commitUpdates();
        }
    }
```

Finally, call the Activity method finish( ) from the onClick handler. This will clean up and close your Activity. The finished NameEditor.java file should look like this:

```
package android_programmers_guide.FindAFriend;

import android_programmers_guide.FindAFriend.Friends;
import android.app.Activity;
import android.database.Cursor;
import android.net.Uri;
import android.os.Bundle;
import android.view.View;
import android.widget.Button;
import android.widget.EditText;


public class NameEditor extends Activity implements View.OnClickListener {

    public static final String EDIT_NAME_ACTION =
        "android_programmers_guide.FindAFriend.action.EDIT_NAME";

    private static final int NAME_INDEX = 1;

    private static final String[] PROJECTION = new String[] {
            Friends.Friend._ID,
```

```
            Friends.Friend.NAME,
    };

    Cursor mCursor;
    EditText mText;

    @Override
    public void onCreate(Bundle icicle) {
        super.onCreate(icicle);

        setContentView(R.layout.name_editor);

        Uri uri = getIntent().getData();

        mCursor = managedQuery(uri, PROJECTION, null, null);

        mText = (EditText) this.findViewById(R.id.name);
        mText.setOnClickListener(this);

        Button b = (Button) findViewById(R.id.ok);
        b.setOnClickListener(this);
    }

    @Override
    protected void onResume() {
        super.onResume();

        if (mCursor != null) {
            mCursor.first();
            String title = mCursor.getString(NAME_INDEX);
            mText.setText(title);
        }
}

    @Override
    protected void onPause() {
        super.onPause();

        if (mCursor != null) {
            String title = mText.getText().toString();
            mCursor.updateString(NAME_INDEX, title);
            mCursor.commitUpdates();
        }
    }

    public void onClick(View v) {
         finish();
    }
}
```

At this point, you can edit name values in the Friends database. However, there are two fields of importance in the database, name and location. In the next section, you will create an editor for the location field.

Creating the LocationEditor Activity

In this section, you will create an editor for the location field of the Friends database. You are going to make this Activity slightly different from the NameEditor Activity. Therefore, the code will be different and follow a slightly unfamiliar process.

If you explored the Google demo NotePad, you should have noticed that the "notes" editor is a white screen with a dynamically drawn line on it that repeats itself as needed. This effect is performed using a custom View. You are going to use this same custom View for the LocationEditor.

location_editor.xml

The first step is to create location_editor.xml and LocationEditor.java files for the layout and code, respectively. The layout file should contain a call to the custom View layout. The full layout is as follows:

```
<?xml version="1.0" encoding="utf-8"?>
<view xmlns:android="http://schemas.android.com/apk/res/android"
class="android_programmers_guide.FindAFriend.LocationEditor$MyEditText"
android:id="@+id/location"
    android:layout_width="fill_parent"
    android:layout_height="fill_parent"
    android:background="#ffffff"
    android:padding="10dip"
    android:scrollbars="vertical"
    android:fadingEdge="vertical" />
```

The LocationEditor will also contain a menu system that will allow the user to discard, delete, or revert any changes they make. This will be a pretty complex Activity. Therefore, it is best to start at the beginning, the imports section of the LocationEditor.java.

LocationEditor.java

Take a look at the following imports for this Activity, many of which deal with drawing the custom View on the screen:

```
import android.app.Activity;
import android.content.ComponentName;
```

```
import android.content.Context;
import android.content.Intent;
import android.database.Cursor;
import android.graphics.Canvas;
import android.graphics.Paint;
import android.graphics.Rect;
import android.net.Uri;
import android.os.Bundle;
import android.util.AttributeSet;
import android.view.Menu;
import android.widget.EditText;
import java.util.Map;
```

Next, set up your Activity's main class outline. There are a number of variables that you need to define for use throughout the LocationEditor:

```
public class LocationEditor extends Activity {

    private static final String TAG = "Friends";

    private static final int FRIEND_INDEX = 1;
    private static final int NAME_INDEX = 2;
    private static final int MODIFIED_INDEX = 3;

    private static final String[] PROJECTION = new String[] {
            Friends.Friend._ID, // 0
            Friends.Friend.LOCATION, // 1
            Friends.Friend.NAME, // 2
            Friends.Friend.MODIFIED_DATE // 3
    };

    private static final String ORIGINAL_CONTENT = "origContent";

    private static final int REVERT_ID = Menu.FIRST;
    private static final int DISCARD_ID = Menu.FIRST + 1;
    private static final int DELETE_ID = Menu.FIRST + 2;

    private static final int STATE_EDIT = 0;
    private static final int STATE_INSERT = 1;

    private int mState;
    private boolean mNoteOnly = false;
    private Uri mURI;
    private Cursor mCursor;
```

```
    private EditText mText;
    private String mOriginalContent;
}
```

Having performed the tasks in the previous sections of this chapter, the variable definitions here should be rather self-explanatory.

The next piece of code shows a subclass that you need to create. This subclass will draw to the screen the EditText that will be used for the LocationEditor. You are breaking this out so that you can call it as needed from the Activity. Keep in mind that you will dynamically draw a new EditText on the screen as is needed by the user. Pay special attention to the onDraw class that you need to override.

```
public static class MyEditText extends EditText {
    private Rect mRect;
    private Paint mPaint;

    public MyEditText(Context context, AttributeSet attrs, Map params) {
        super(context, attrs, params);

        mRect = new Rect();
        mPaint = new Paint();
        mPaint.setStyle(Paint.Style.STROKE);
        mPaint.setColor(0xFF0000FF);
    }
    @Override
    protected void onDraw(Canvas canvas) {

        int count = getLineCount();
        Rect r = mRect;
        Paint paint = mPaint;

        for (int i = 0; i < count; i++) {
            int baseline = getLineBounds(i, r);

            canvas.drawLine(r.left, baseline + 1, r.right, baseline + 1,
                            paint);
        }

        super.onDraw(canvas);
    }
}
```

Again, while this may look like a lot of code, there should be nothing foreign about it. This subclass simply draws a new EditText as needed to the screen.

Just as with the NameEditor, you will use the onResume( ) and onPause( ) methods to do your database work. Take a look at the code for each below:

```
protected void onResume() {
     super.onResume();

     if (mCursor != null) {
         mCursor.first();

         if (mState == STATE_EDIT) {
             setTitle(getText(R.string.title_edit));
         } else if (mState == STATE_INSERT) {
             setTitle(getText(R.string.title_create));
         }

         String note = mCursor.getString(FRIEND_INDEX);
         mText.setTextKeepState(note);

         if (mOriginalContent == null) {
             mOriginalContent = note;
         }

     } else {
         setTitle(getText(R.string.error_title));
         mText.setText(getText(R.string.error_message));
     }
}
protected void onPause() {
     super.onPause();

      if (mCursor != null) {
         String text = mText.getText().toString();
         int length = text.length();

         if (isFinishing() && (length == 0) && !mNoteOnly) {
             setResult(RESULT_CANCELED);
             deleteFriend();
         } else {
             if (!mNoteOnly) {
                 mCursor.updateLong(MODIFIED_INDEX,
System.currentTimeMillis());

                 if (mState == STATE_INSERT) {
                     String title = text.substring(0, Math.min(30,
length));
```

```
                    if (length > 30) {
                        int lastSpace = title.lastIndexOf(' ');
                        if (lastSpace > 0) {
                            title = title.substring(0, lastSpace);
                        }
                    }
                    mCursor.updateString(NAME_INDEX, title);
                }
            }

            mCursor.updateString(FRIEND_INDEX, text);

            managedCommitUpdates(mCursor);
        }
    }
}
```

Much like in the NameEditor, you read from the database during onResume( ) and write back to it during onPause( ). The one added feature that appears in LocationEditor as opposed to NameEditor is that you are also writing out the modified dates when you make a change.

Finally, you need two methods for canceling changes and deleting friends. These methods will be called from the menu system:

```
private final void cancelFriend() {
    if (mCursor != null) {
        if (mState == STATE_EDIT) {
            mCursor.updateString(FRIEND_INDEX, mOriginalContent);
            mCursor.commitUpdates();
            mCursor.deactivate();
            mCursor = null;
        } else if (mState == STATE_INSERT) {
            deleteFriend();
        }
    }
    setResult(RESULT_CANCELED);
    finish();
}

private final void deleteFriend() {
    if (mCursor != null) {
        mText.setText("");
        mCursor.deleteRow();
        mCursor.deactivate();
```

```
            mCursor = null;
        }
    }
```

Given that you learned about creating menu systems in Chapter 8, simply examine the full LocationEditor.java file to see how all of these methods and subclasses work together:

```
package android_programmers_guide.FindAFriend;


import android.app.Activity;
import android.content.ComponentName;
import android.content.Context;
import android.content.Intent;
import android.database.Cursor;
import android.graphics.Canvas;
import android.graphics.Paint;
import android.graphics.Rect;
import android.net.Uri;
import android.os.Bundle;
import android.util.AttributeSet;
import android.view.Menu;
import android.widget.EditText;
import java.util.Map;

public class LocationEditor extends Activity {

    private static final String TAG = "Friends";

    private static final int FRIEND_INDEX = 1;
    private static final int NAME_INDEX = 2;
    private static final int MODIFIED_INDEX = 3;

    private static final String[] PROJECTION = new String[] {
            Friends.Friend._ID, // 0
            Friends.Friend.LOCATION, // 1
            Friends.Friend.NAME, // 2
            Friends.Friend.MODIFIED_DATE // 3
    };

    private static final String ORIGINAL_CONTENT = "origContent";

    private static final int REVERT_ID = Menu.FIRST;
    private static final int DISCARD_ID = Menu.FIRST + 1;
    private static final int DELETE_ID = Menu.FIRST + 2;

    private static final int STATE_EDIT = 0;
    private static final int STATE_INSERT = 1;

    private int mState;
```

```
private boolean mNoteOnly = false;
private Uri mURI;
private Cursor mCursor;
private EditText mText;
private String mOriginalContent;

public static class MyEditText extends EditText {
    private Rect mRect;
    private Paint mPaint;

    public MyEditText(Context context, AttributeSet attrs, Map params) {
        super(context, attrs, params);

        mRect = new Rect();
        mPaint = new Paint();
        mPaint.setStyle(Paint.Style.STROKE);
        mPaint.setColor(0xFF0000FF);
    }

    @Override
    protected void onDraw(Canvas canvas) {

        int count = getLineCount();
        Rect r = mRect;
        Paint paint = mPaint;

        for (int i = 0; i < count; i++) {
            int baseline = getLineBounds(i, r);

            canvas.drawLine(r.left, baseline + 1, r.right, baseline + 1,
                    paint);
        }

        super.onDraw(canvas);
    }
}

@Override
protected void onCreate(Bundle icicle) {
    super.onCreate(icicle);

    final Intent intent = getIntent();
    final String type = intent.resolveType(this);

     final String action = intent.getAction();
    if (action.equals(Intent.EDIT_ACTION)) {
        mState = STATE_EDIT;
        mURI = intent.getData();

    } else if (action.equals(Intent.INSERT_ACTION)) {
        mState = STATE_INSERT;
```

```
            mURI = getContentResolver().insert(intent.getData(), null);

            if (mURI == null) {


                finish();
                return;
            }

            setResult(RESULT_OK, mURI.toString());

        } else {
          finish();
            return;
        }

        setContentView(R.layout.location_editor);

        mText = (EditText) findViewById(R.id.location);

        mCursor = managedQuery(mURI, PROJECTION, null, null);

        if (icicle != null) {
            mOriginalContent = icicle.getString(ORIGINAL_CONTENT);
        }
    }


    @Override
    protected void onResume() {
        super.onResume();

        if (mCursor != null) {
            mCursor.first();

            if (mState == STATE_EDIT) {
                setTitle(getText(R.string.title_edit));
            } else if (mState == STATE_INSERT) {
                setTitle(getText(R.string.title_create));
            }

            String note = mCursor.getString(FRIEND_INDEX);
            mText.setTextKeepState(note);

            if (mOriginalContent == null) {
                mOriginalContent = note;
            }

        } else {
            setTitle(getText(R.string.error_title));
            mText.setText(getText(R.string.error_message));
        }
    }
```

```
    @Override
    protected void onFreeze(Bundle outState) {
        outState.putString(ORIGINAL_CONTENT, mOriginalContent);
    }

    @Override
    protected void onPause() {
        super.onPause();

         if (mCursor != null) {
            String text = mText.getText().toString();
            int length = text.length();

            if (isFinishing() && (length == 0) && !mNoteOnly) {
                setResult(RESULT_CANCELED);
                deleteFriend();
            } else {
                if (!mNoteOnly) {
                    mCursor.updateLong(MODIFIED_INDEX,
System.currentTimeMillis());

                    if (mState == STATE_INSERT) {
                        String title = text.substring(0, Math.min(30,
length));
                        if (length > 30) {
                            int lastSpace = title.lastIndexOf(' ');
                            if (lastSpace > 0) {
                                title = title.substring(0, lastSpace);
                            }
                        }
                        mCursor.updateString(NAME_INDEX, title);
                    }
                }

                mCursor.updateString(FRIEND_INDEX, text);

                managedCommitUpdates(mCursor);
            }
        }
    }

    @Override
    public boolean onCreateOptionsMenu(Menu menu) {
        super.onCreateOptionsMenu(menu);

        if (mState == STATE_EDIT) {
            menu.add(0, REVERT_ID, R.string.menu_revert).setShortcut('0',
'r');
            if (!mNoteOnly) {
                menu.add(0, DELETE_ID,
R.string.menu_delete).setShortcut('1', 'd');
            }
```

```
        } else {
            menu.add(0, DISCARD_ID, R.string.menu_discard).setShortcut('0',
'd');
        }

        if (!mNoteOnly) {
            Intent intent = new Intent(null, getIntent().getData());
            intent.addCategory(Intent.ALTERNATIVE_CATEGORY);
            menu.addIntentOptions(
                Menu.ALTERNATIVE, 0,
                new ComponentName(this, LocationEditor.class), null,
                intent, 0, null);
        }

        return true;
    }

    @Override
    public boolean onOptionsItemSelected(Menu.Item item) {
        switch (item.getId()) {
        case DELETE_ID:
            deleteFriend();
            finish();
            break;
        case DISCARD_ID:
            cancelFriend();
            break;
        case REVERT_ID:
            cancelFriend();
            break;
        }
        return super.onOptionsItemSelected(item);
    }

    private final void cancelFriend() {
        if (mCursor != null) {
            if (mState == STATE_EDIT) {
                mCursor.updateString(FRIEND_INDEX, mOriginalContent);
                mCursor.commitUpdates();
                mCursor.deactivate();
                mCursor = null;
            } else if (mState == STATE_INSERT) {
                deleteFriend();
            }
        }
        setResult(RESULT_CANCELED);
        finish();
    }

    private final void deleteFriend() {
        if (mCursor != null) {
            mText.setText("");
```

```
                mCursor.deleteRow();
                mCursor.deactivate();
                mCursor = null;
            }
        }
}
```

In the next section, you will create the Activity that will draw your Google Maps Overlay. The FriendsMap activity will read the full recordset of friends from the Friends database and write each to the Overlay.

Creating the FriendsMap Activity

The FriendsMap Activity is the final Activity that will be callable from the main application. This Activity will call your recordset from the Friends database and draw a circle on a Google Map for each friend. The Activity will also draw a circle for you at your current location.

You need to begin by adding two new files to your project, friendsmap.xml and FriendsMap.java. Because you have seen the layout for the friendsmap.xml file in Chapter 9, there is no need to fully explain it here. You are using a RelativeLayout to place four Buttons over a Google Map. The full friendsmap.xml file should look like this:

```
<?xml version="1.0" encoding="utf-8"?>
<RelativeLayout xmlns:android="http://schemas.android.com/apk/res/android"
    android:orientation="vertical"
    android:layout_width="fill_parent"
    android:layout_height="fill_parent"
    >
  <view class="com.google.android.maps.MapView"
        android:id="@+id/myMap"
        android:layout_width="wrap_content"
        android:layout_height="wrap_content"/>
          <Button android:id="@+id/buttonZoomIn"
            style="?android:attr/buttonStyleSmall"
            android:text="+"
            android:layout_width="wrap_content"
            android:layout_height="wrap_content" />
          <Button android:id="@+id/buttonMapView"
            style="?android:attr/buttonStyleSmall"
            android:text="Map"
            android:layout_alignRight="@+id/myMap"
            android:layout_width="wrap_content"
            android:layout_height="wrap_content" />
          <Button android:id="@+id/buttonSatView"
            style="?android:attr/buttonStyleSmall"
            android:text="Sat"
            android:layout_alignRight="@+id/myMap"
```

```
            android:layout_alignBottom="@+id/myMap"
            android:layout_width="wrap_content"
            android:layout_height="wrap_content" />
          <Button android:id="@+id/buttonZoomOut"
            style="?android:attr/buttonStyleSmall"
            android:text="-"
            android:layout_alignBottom="@+id/myMap"
            android:layout_width="wrap_content"
            android:layout_height="wrap_content" />
</RelativeLayout>
```

Because you have seen the overwhelming majority of the FriendsMap.java file in Chapter 9 as well, I will not go over every little detail. However, there is one method that should be explained.

You will create a method called LoadFriends( ) that will access the database, read the records, and draw the Overlay. Take a look at the LoadFriends( ) code that follows. Notice that you open the database, match and parse the location field, create a point from the latitude and longitude in the location field, and draw that point to the Overlay. The last thing the method does is to grab your coordinates from the GPS and draw them to the Overlay with the label "ME".

```
    public void LoadFriends(MapView mv, MapController mc, Cursor c){
      Point myLocation = null;
      Double latPoint = null;
      Double lngPoint = null;
      c.first();
      do{
            if (c.getString(c.getColumnIndex("location")) != null) {
           final String geoPattern = "(geo:[\\-]?[0-9]{1,3}\\.[0
9]{1,6}\\,[\\-]?[0-9]{1,3}\\.[0-9]{1,6}\\#)";
                 Pattern pattern = Pattern.compile(geoPattern);

                 CharSequence inputStr =
c.getString(c.getColumnIndex("location"));
                 Matcher matcher = Pattern.matcher(inputStr);

                 boolean matchFound = matcher.find();
                 if (matchFound) {
                      String groupStr = matcher.group(0);
                      latPoint =
Double.valueOf(groupStr.substring(groupStr.indexOf(":") + 1,
                                 groupStr.indexOf(","))) ;
                      lngPoint =
Double.valueOf(groupStr.substring(groupStr.indexOf(",") + 1,
                                 groupStr.indexOf("#"))) ;
                      Point friendLocation = new
Point(latPoint.intValue(),lngPoint.intValue());
```

```
drawFriendsOverlay.addNewFriend(c.getString(c.getColumnIndex("name")),
friendLocation);
                  }
              }
       }while(c.next());
       LocationManager myManager = (LocationManager)
getSystemService(Context.LOCATION_SERVICE);
       Double myLatPoint =
myManager.getCurrentLocation("gps").getLatitude()*1E6;
         Double myLngPoint =
myManager.getCurrentLocation("gps").getLongitude()*1E6;
       myLocation = new Point(myLatPoint.intValue(),myLngPoint.intValue());
         drawFriendsOverlay.addNewFriend("Me", myLocation);

       mc.centerMapTo(myLocation, false);
       mc.zoomTo(9);
        mv = null;
    }
```

The remainder of the FriendsMap.java file operates the zoom and toggle buttons, as introduced in Chapter 10:

```
package android_programmers_guide.FindAFriend;


import android.os.Bundle;
import android.location.LocationManager;
import android.view.View;
import android.content.Context;
import android.content.Intent;
import android.database.Cursor;
import android.widget.Button;
import java.util.regex.Pattern;
import java.util.regex.Matcher;
import android.graphics.Canvas;
import android.graphics.RectF;
import android.graphics.Paint;
import com.google.android.maps.MapActivity;
import com.google.android.maps.MapView;
import com.google.android.maps.Point;
import com.google.android.maps.MapController;
import com.google.android.maps.Overlay;
import com.google.android.maps.OverlayController;

public class FriendsMap extends MapActivity {


    private static final String[] PROJECTION = new String[] {
        Friends.Friend.NAME, Friends.Friend.LOCATION};
    public  Cursor mCursor;
```

```
    DrawFriendsOverlay drawFriendsOverlay = new DrawFriendsOverlay();

    @Override
    public void onCreate(Bundle icicle) {
        super.onCreate(icicle);
        setContentView(R.layout.friendsmap);

        Intent intent = getIntent();
        if (intent.getData() == null) {
            intent.setData(Friends.Friend.CONTENT_URI);
        }
        mCursor = managedQuery(getIntent().getData(), PROJECTION, null,null);

        final MapView myMap = (MapView) findViewById(R.id.myMap);
        final MapController myMapController = myMap.getController();
        LoadFriends(myMap, myMapController, mCursor);
        OverlayController myOverlayController =
myMap.createOverlayController();
        myOverlayController.add(drawFriendsOverlay, true);
        final Button zoomIn = (Button) findViewById(R.id.buttonZoomIn);
                zoomIn.setOnClickListener(new Button.OnClickListener() {
              public void onClick(View v){
                    ZoomIn(myMap,myMapController);
              }});
        final Button zoomOut = (Button) findViewById(R.id.buttonZoomOut);
        zoomOut.setOnClickListener(new Button.OnClickListener() {
              public void onClick(View v){
                    ZoomOut(myMap,myMapController);
              }});
       final Button viewMap = (Button) findViewById(R.id.buttonMapView);
       viewMap.setOnClickListener(new Button.OnClickListener() {
              public void onClick(View v){
                    ShowMap(myMap,myMapController);
              }});
       final Button viewSat = (Button) findViewById(R.id.buttonSatView);
       viewSat.setOnClickListener(new Button.OnClickListener() {
              public void onClick(View v){
                    ShowSat(myMap,myMapController);
              }});

    }

    public void LoadFriends(MapView mv, MapController mc, Cursor c){
      Point myLocation = null;
      Double latPoint = null;
      Double lngPoint = null;
      c.first();
      do{
            if (c.getString(c.getColumnIndex("location")) != null) {
           final String geoPattern = "(geo:[\\-]?[0-9]{1,3}\\.[0
9]{1,6}\\,[\\-]?[0-9]{1,3}\\.[0-9]{1,6}\\#)";
                  Pattern pattern = Pattern.compile(geoPattern);
```

```
                CharSequence inputStr =
c.getString(c.getColumnIndex("location"));
                Matcher matcher = pattern.matcher(inputStr);

                boolean matchFound = matcher.find();
                if (matchFound) {
                     String groupStr = matcher.group(0);
                     latPoint =
Double.valueOf(groupStr.substring(groupStr.indexOf(":") + 1,
                               groupStr.indexOf(","))) ;
                     lngPoint =
Double.valueOf(groupStr.substring(groupStr.indexOf(",") + 1,
                               groupStr.indexOf("#"))) ;
                     Point friendLocation = new
Point(latPoint.intValue(),lngPoint.intValue());

drawFriendsOverlay.addNewFriend(c.getString(c.getColumnIndex("name")),
friendLocation);
                }
           }
      }while(c.next());
      LocationManager myManager = (LocationManager)
getSystemService(Context.LOCATION_SERVICE);
      Double myLatPoint =
myManager.getCurrentLocation("gps").getLatitude()*1E6;
        Double myLngPoint =
myManager.getCurrentLocation("gps").getLongitude()*1E6;
      myLocation = new Point(myLatPoint.intValue(),myLngPoint.intValue());
        drawFriendsOverlay.addNewFriend("Me", myLocation);

      mc.centerMapTo(myLocation, false);
      mc.zoomTo(9);
       mv = null;
    }

    public void ZoomIn(MapView mv, MapController mc){
      if(mv.getZoomLevel()!=21){
      mc.zoomTo(mv.getZoomLevel()+ 1);
      }
    }
    public void ZoomOut(MapView mv, MapController mc){
      if(mv.getZoomLevel()!=1){
            mc.zoomTo(mv.getZoomLevel()- 1);
            }
    }
    public void ShowMap(MapView mv, MapController mc){
            if (mv.isSatellite()){
                  mv.toggleSatellite();
            }
    }
    public void ShowSat(MapView mv, MapController mc){
            if (!mv.isSatellite()){
```

```
                mv.toggleSatellite();
            }
    }
    protected class DrawFriendsOverlay extends Overlay{
      public String[] friendName = new String[0];
      public Point[] friendPoint = new Point[0];
      final Paint paint = new Paint();

      @Override
      public void draw(Canvas canvas, PixelCalculator calculator, Boolean
shadow){
            for(int x=0;x<friendPoint.length; x++){
                  int[] coords = new int[2];
                  calculator.getPointXY(friendPoint[x], coords);
          RectF oval = new RectF(coords[0] - 7, coords[1] + 7,
                          coords[0] + 7, coords[1] - 7);
          paint.setTextSize(14);
            canvas.drawText(friendName[x],
                      coords[0] +9, coords[1], paint);
            canvas.drawOval(oval, paint);

             }
      }
      public void addNewFriend(String name,Point point ){
      int x = friendPoint.length;

            String[] friendNameB = new String[x + 1];
            Point[] friendPointB = new Point[x + 1];

            System.arraycopy(friendName, 0, friendNameB, 0, x );
            System.arraycopy(friendPoint, 0, friendPointB, 0, x);

            friendNameB[x] = name;
            friendPointB[x]= point;

            friendName = new String[x + 1];
            friendPoint = new Point[x + 1];
            System.arraycopy(friendNameB, 0, friendName, 0, x + 1 );
            System.arraycopy(friendPointB, 0, friendPoint, 0, x + 1 );


      }

    }
}
```

The last task to finish this project is to create the main Activity, FindAFriend, which will be a shell that calls the other Activities you created in this chapter.

Creating the FindAFriend Activity

To begin this section, create two files, findafriend.xml and FindAFriend.java. Once again, these files will hold your layout and code for the current section, respectively.

The layout file is very basic and contains only a TextView. This TextView will be used to write results to in your list of friends. The full findafriend.xml file should appear as follows:

```
<?xml version="1.0" encoding="utf-8"?>
<TextView xmlns:android="http://schemas.android.com/apk/res/android"
    android:id="@android:id/text1"
    android:layout_width="fill_parent"
    android:layout_height="?android:attr/listPreferredItemHeight"
    android:textAppearance="?android:attr/textAppearanceLargeInverse"
    android:gravity="center_vertical"
    android:paddingLeft="27dip"
/>
```

The full contents of the FindAFriend.java file follows. All of the code in this file has already been covered in this chapter. First, you read the database and write the results to a ListView. The user is then given menu options to edit or delete entries, or launch the FriendsMap Activity. Piece of cake, right?

```
package android_programmers_guide.FindAFriend;

import android_programmers_guide.FindAFriend.Friends;
import android.app.ListActivity;
import android.content.ComponentName;
import android.content.Intent;
import android.content.ContentUris;
import android.database.Cursor;
import android.graphics.Color;
import android.net.Uri;
import android.os.Bundle;
import android.view.Menu;
import android.view.View;
import android.view.View.MeasureSpec;
import android.widget.ListAdapter;
import android.widget.ListView;
import android.widget.SimpleCursorAdapter;
import android.widget.TextView;

public class FindAFriend extends ListActivity {
```

```
    public static final int DELETE_ID = Menu.FIRST;
    public static final int INSERT_ID = Menu.FIRST + 1;
    public static final int FIND_FRIENDS = Menu.FIRST + 2;

    private static final String[] PROJECTION = new String[] {
            Friends.Friend._ID, Friends.Friend.NAME};

     private Cursor mCursor;

    @Override
    protected void onCreate(Bundle icicle) {
        super.onCreate(icicle);

        setDefaultKeyMode(SHORTCUT_DEFAULT_KEYS);

        Intent intent = getIntent();
        if (intent.getData() == null) {
            intent.setData(Friends.Friend.CONTENT_URI);
        }

        setupList();

        mCursor = managedQuery(getIntent().getData(), PROJECTION, null,
null);

        ListAdapter adapter = new SimpleCursorAdapter(this,
                R.layout.findafriend_item, mCursor,
                new String[] {Friends.Friend.NAME}, new int[]
{android.R.id.text1});
        setListAdapter(adapter);

    }

    private void setupList() {
        View view = getViewInflate().inflate(
                android.R.layout.simple_list_item_1, null, null);

        TextView v = (TextView) view.findViewById(android.R.id.text1);
        v.setText("X");
        getListView().setBackgroundColor(Color.GRAY);
        v.measure(MeasureSpec.makeMeasureSpec(View.MeasureSpec.EXACTLY,
100),
                MeasureSpec.makeMeasureSpec(View.MeasureSpec.UNSPECIFIED,
0));
    }

    @Override
    public boolean onCreateOptionsMenu(Menu menu) {
        super.onCreateOptionsMenu(menu);
```

```
        menu.add(0, INSERT_ID, R.string.menu_insert).setShortcut('3', 'a');

        Intent intent = new Intent(null, getIntent().getData());
        intent.addCategory(Intent.ALTERNATIVE_CATEGORY);
        menu.addIntentOptions(
            Menu.ALTERNATIVE, 0, new ComponentName(this, FindAFriend.class),
            null, intent, 0, null);

        return true;
    }

    @Override
    public boolean onPrepareOptionsMenu(Menu menu) {
        super.onPrepareOptionsMenu(menu);
        final boolean haveItems = mCursor.count() > 0;

        if (haveItems) {
            Uri uri = ContentUris.withAppendedId(getIntent().getData(),
getSelectedItemId());

            Intent[] specifics = new Intent[1];
            specifics[0] = new Intent(Intent.EDIT_ACTION, uri);
            Menu.Item[] items = new Menu.Item[1];

            Intent intent = new Intent(null, uri);
            intent.addCategory(Intent.SELECTED_ALTERNATIVE_CATEGORY);
            menu.addIntentOptions(Menu.SELECTED_ALTERNATIVE, 0, null,
specifics, intent, 0, items);
            menu.add(Menu.SELECTED_ALTERNATIVE, DELETE_ID,
R.string.menu_delete)
                    .setShortcut('2', 'd');
            menu.add(Menu.SELECTED_ALTERNATIVE, FIND_FRIENDS,
R.string.find_friends).setShortcut('4', 'f');
            if (items[0] != null) {
                items[0].setShortcut('1', 'e');
            }
        } else {
            menu.removeGroup(Menu.SELECTED_ALTERNATIVE);
        }

        menu.setItemShown(DELETE_ID, haveItems);
        return true;
    }

    @Override
    public boolean onOptionsItemSelected(Menu.Item item) {
        switch (item.getId()) {
        case DELETE_ID:
            deleteItem();
```

```
                return true;
            case INSERT_ID:
                insertItem();
                return true;
            case FIND_FRIENDS:
                Intent findfriends = new Intent(this, FriendsMap.class);
                startActivity(findfriends);
                return true;
            }
            return super.onOptionsItemSelected(item);
        }

        @Override
        protected void onListItemClick(ListView l, View v, int position, long
id) {
            Uri url = ContentUris.withAppendedId(getIntent().getData(), id);

            String action = getIntent().getAction();
            if (Intent.PICK_ACTION.equals(action)
                    || Intent.GET_CONTENT_ACTION.equals(action)) {
                setResult(RESULT_OK, url.toString());
            } else {
                startActivity(new Intent(Intent.EDIT_ACTION, url));
            }
        }

        private final void deleteItem() {
            mCursor.moveTo(getSelectedItemPosition());
            mCursor.deleteRow();
        }
        private final void insertItem() {
            startActivity(new Intent(Intent.INSERT_ACTION,
getIntent().getData()));
        }
}
```

While this was the longest Activity that you have created in this book, it should still be noted that the relative amount of programming needed to do what you did is fairly small. Next, run this Activity and see the result of all your work.

Running the FindAFriend Activity

Run the FindAFriend Activity in the Android Emulator. You should be greeted with an empty list, as shown in the following illustration. To add your first friend, click the Menu button and select the Add Friend option.

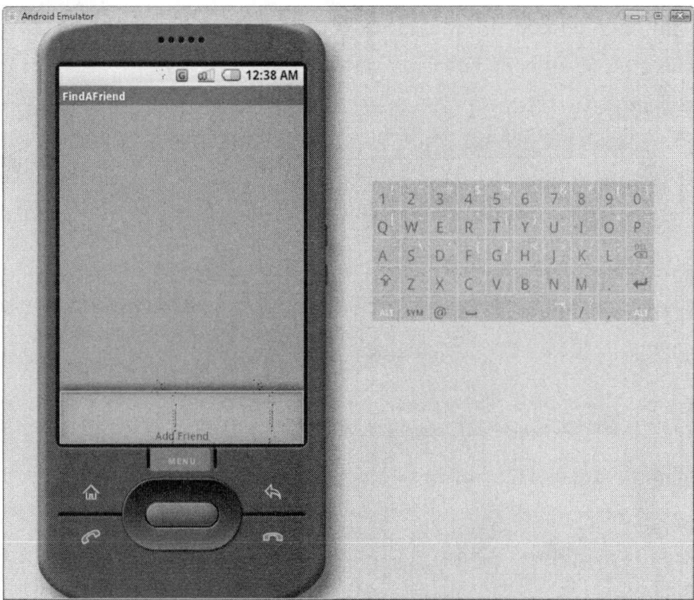

This option launches the custom View you created. Enter a friend's name on the line provided, as shown here, and return to the main Activity by clicking the back arrow on the Emulator.

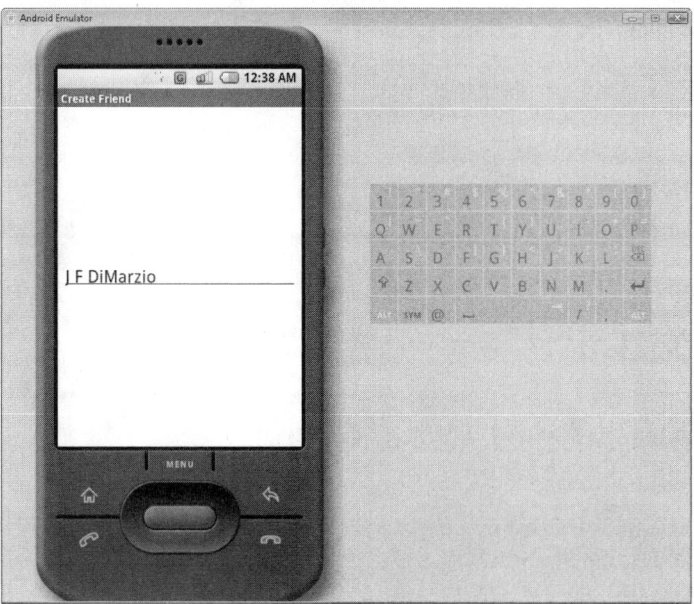

You should now have a friend's name in the ListView. Click the Menu button again; you clearly have more options now, as shown in this illustration.

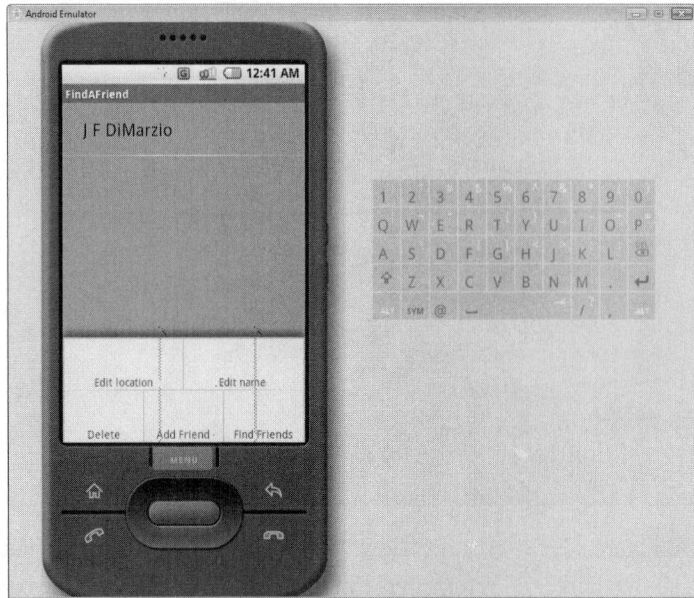

Select the Edit Location option. This should bring up your custom control yet again. Enter a coordinate-based location, as shown here.

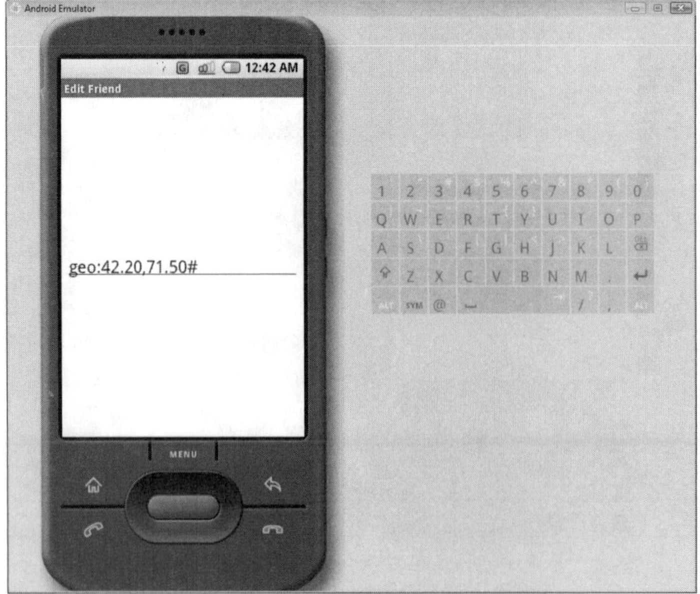

Finally, return to the main Activity and select the Find Friends option. This should clearly map out your current location in San Francisco and your friend's location off the coast of Africa, respectively.

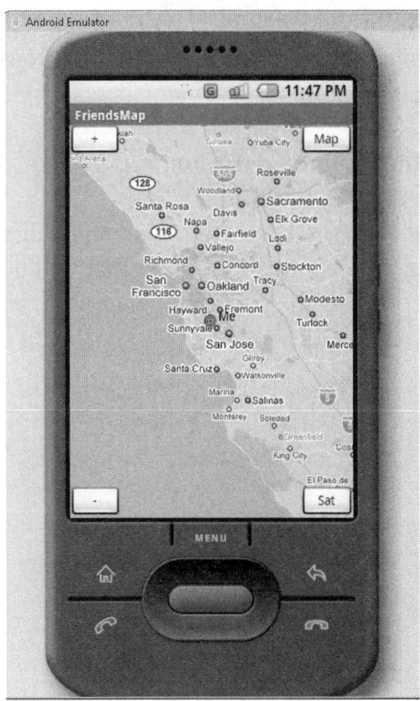

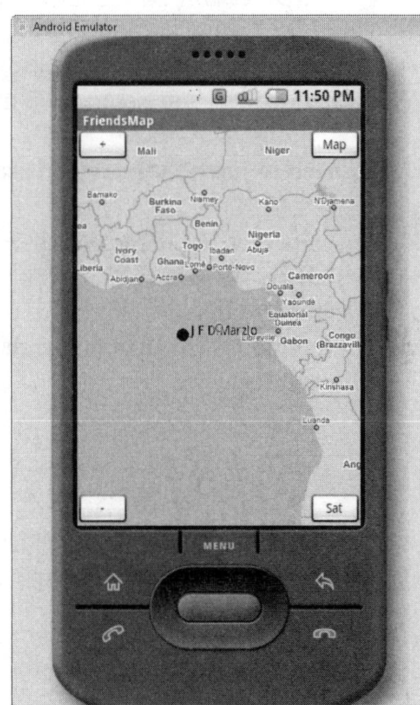

Try This Real-Time Location Updating

Try modifying the FindAFriend application to update the "ME" marker as you move. This should be fairly easy to do using the update( ) method.

Chapter 12, the final chapter, provides a reference to some of the Android SDK options, such as the adb commands and the Android Emulator options.

Ask the Expert

Q: Can a SQLite database be created in code?

A: Yes. However, for the purposes of giving a well-rounded tutorial on Android, I chose to give an example of manually creating the database. Feel free to modify this project to include an in-code database-creation method in the FriendsProvider Content Provider.

Q: Do you need to have a separate class that implements BaseColumns?

A: No. You can define the items from the Friends class (in this chapter's example) directly in the calling class. However, if you are creating a Content Provider that will be implemented by other developers who may not know the underlying data structure, you will want to provide a defining class.

Chapter 12

Android SDK Tool Reference

This chapter provides a valuable reference to some of the Android SDK tools that you have used over the course of this book. It gives you some of the command-line options that you can use with the Android Emulator and the Android Debugging Bridge.

Android Emulator Commands

The following table contains a list of the most common Android Emulator commands. These are the commands that were available as of the March 2008 SDK release. A short description is provided with each command.

Emulator Command	Function
`emulator -console`	Enables the console shell on the current terminal
`emulator -data <filename>`	Uses a different file as the working user-data disk image
`emulator -debug-kernel`	Sends kernel output to the console
`emulator -flash-keys`	Flashes keypresses on the device skin
`emulator -help`	Prints a list of all Emulator commands
`emulator -http-proxy <proxy>`	Makes all TCP connections through a specified HTTP/HTTPS proxy
`emulator -image <file>`	Uses <file> as the system image
`emulator -kernel <file>`	Uses <file> as the emulated kernel
`emulator -logcat <logtags>`	Enables logcat output with given tags
`emulator -mic <device or file>`	Uses device or WAV file for audio input
`emulator -netdelay <delay>`	Sets network latency emulation to <delay>. (The <delay> parameter simulates the delay experienced on specific types of networks.) The <delay>s you can use are as follows: • Gprs • Edge • Umts • None • <num> • <min>:<max>
`emulator -netfast`	Shortcut for `-netspeed full -netdelay none`

Emulator Command	Function
`emulator -netspeed <speed>`	Sets network speed emulation to <speed>. (The <speed> parameter simulates the data speed experienced on specific types of networks.) The <speed>s you can use are as follows: • Gsm • Hscsd • Gprs • Edge • Umts • Hsdpa • Full • <num> • <up>:<down>
`emulator -noaudio`	Disables Android audio support
`emulator -nojni`	Disables JNI checks in the Dalvik virtual machine
`emulator -noskin`	Specifies not to use any Emulator skin
`emulator -onion <image>`	Uses overlay image over screen
`emulator -onion-alpha <percent>`	Specifies onion skin translucency value (as percent)
`emulator -qemu`	Passes arguments to QEMU
`emulator -qemu -h`	Displays QEMU help
`emulator -radio <device>`	Redirects the radio modem interface to a host character device
`emulator -ramdisk <file>`	Uses <file> as the ramdisk image
`emulator -raw-keys`	Disables Unicode keyboard reverse mapping
`emulator -sdcard <file>`	Uses <file> as the SD Memory Card image
`emulator -skin <skinID>`	Starts the Emulator with the specified skin: • HVGA-L 480x320, landscape • HVGA-P 320x480, portrait (default) • QVGA-L 320x240, landscape • QVGA-P 240x320, portrait
`emulator -skindir <dir>`	Searches for Emulator skins in <dir>
`emulator -system <dir>`	Searches system, ramdisk, and user-data disk images in <dir>
`emulator -trace <name>`	Enables code profiling (press F9 to start), written to a specified file

Emulator Command	Function
`emulator -useaudio`	Enables Android audio support
`emulator -verbose`	Enables verbose output
`emulator -verbose-keys`	Enables verbose keypress messages
`emulator -verbose-proxy`	Enables verbose proxy debug messages
`emulator -wipe-data`	Deletes all data on the user-data disk image (see `emulator -data <filename>`) before starting

Android Debug Bridge Commands

The following commands are gsm commands. You access them by connecting to the Emulator's terminal console. If you do not know the port terminal console, it is one less than the debug port. Execute adb devices to get a list of active devices and the related port numbers.

adb Command	Function
`adb Bugreport`	Prints dumpsys, dumpstate, and logcat data to the screen, for the purposes of bug reporting
`adb call <phonenumber>`	Simulates an inbound phone call from <phonenumber>
`adb cancel <phonenumber>`	Cancels an inbound phone call from <phonenumber>
`adb -d {<ID>|<serialNumber>}`	Lets you direct an adb command to a specific Emulator/device instance, referred to by its adb-assigned ID or serial number
`adb data <state>`	Changes the state of the GPRS data connection to <state>
`adb Devices`	Prints a list of all attached Emulator/device instances
`adb forward <local> <remote>`	Forwards socket connections from a specified local port to a specified remote port on the Emulator/device instance
`adb get-serialno`	Prints the adb instance identifier string
`adb get-state`	Prints the adb state of an Emulator/device instance

adb Command	Function
adb help	Prints a list of supported adb commands
adb install <path-to-apk>	Pushes an Android application (specified as a full path to an .apk file) to the data file of an Emulator/device
adb jdwp	Prints a list of available JDWP processes on a given device
adb kill-server	Terminates the adb server process
adb logcat [<option>] [<filter-specs>]	Prints log data to the screen
adb ppp <tty> [parm]...	Runs PPP over USB: ● <tty> The tty for PPP stream; for example, dev:/dev/omap_csmi_ttyl ● [parm]... Zero or more PPP/PPPD options, such as defaultroute, local, notty, etc. Note that you should not automatically start a PDP connection.
adb pull <remote> <local>	Copies a specified file from an Emulator/device instance to your development computer
adb push <local> <remote>	Copies a specified file from your development computer to an Emulator/device instance
adb Shell	Starts a remote shell in the target Emulator/device instance
adb start-server	Checks whether the adb server process is running and, if not, starts it
adb Status	Reports the current GSM voice/data state
adb unregistered	Indicates no network is available
adb Version	Prints the adb version number
adb voice <state>	Changes the state of the GPRS voice connection to <state>
adb wait-for-bootloader	Blocks execution until the bootloader is online—that is, until the instance state is bootloader
adb wait-for-device	Blocks execution until the device is online—that is, until the instance state is device

Index

A

B

C

D

H

I

J

K

L

M

N

O

P

Q

R